Bioprinting

Bioprinting

Techniques and Risks for Regenerative Medicine

Maika G. Mitchell
ASCP, NSBE, AACC, CAP, NYSDOH,
Lean Six Sigma Master Black Belt,
MSKCC, CLC bio, The Science Advisory Board,
BioConference Live!, Touch Oncology,
PRIMR, and Lean In Community, e-NABLE

Academic Press is an imprint of Elsevier
125 London Wall, London EC2Y 5AS, United Kingdom
525 B Street, Suite 1800, San Diego, CA 92101-4495, United States
50 Hampshire Street, 5th Floor, Cambridge, MA 02139, United States
The Boulevard, Langford Lane, Kidlington, Oxford OX5 1GB, United Kingdom

British Library Cataloguing-in-Publication Data
A catalogue record for this book is available from the British Library

Library of Congress Cataloging-in-Publication Data
A catalog record for this book is available from the Library of Congress

ISBN: 978-0-12-805369-0

For Information on all Academic Press publications
visit our website at https://www.elsevier.com/books-and-journals

Publisher: Mica Haley
Acquisition Editor: Sara Tenney
Editorial Project Manager: Timothy Bennett
Senior Production Project Manager: Priya Kumaraguruparan

Typeset by MPS Limited, Chennai, India

I first and foremost dedicate this book to my husband Carlton. He first introduced me to the Netflix film, "Print a Legend" while I was writing my first book with Elsevier. This new and incredible world of "makers" changed my life forever.

I thank my children Cody, Carmen, and Grace for being proud of me and encouraging their mother to keep being curious about the world around me.

I also dedicate and thank all those helpdesk engineers from various companies such as New Matter (@new.matter), who skyped with me to fix my 3D printers and bring them back to life.

Lastly, I thank those individuals who told me I couldn't do it which propelled me from a place of no to a place of yes.

CONTENTS

CHAPTER 1

Biomanufacturing: The Definition and Evolution of a New Genre

Biomanufacturing refers to the use of cells or other living microorganisms to produce commercially viable products. Vaccines, monoclonal antibodies, and proteins for medicinal use are all produced by biomanufacturing. Other examples include amino acids, industrial enzymes, biofuels, and biochemicals for consumer and industrial applications. Biomanufacturing is an interdisciplinary field incorporating aspects of chemical engineering, biochemistry, and microbiology.

This chapter provides a brief summary of the developments in biomanufacturing technology, which is very much in its infancy.

WHAT IS 3D PRINTING?

Three-dimensional (3D) printing is a type of additive manufacturing (AM) method whereby objects are created by fusing or depositing materials. Some examples are plastic, metal, ceramics, powders, liquids, or even living cells printed in layers to produce a 3D object [1,2,3]. This process is also referred to as AM, rapid prototyping (RP), or solid free-form technology [4]. Three-dimensional printing is expected to revolutionize medicine and other fields, not unlike the way the printing press transformed publishing [1].

The History of 3D Printing

Charles Hull invented 3D printing, which he called "stereolithography (SLA)," in the early 1980s [1]. Hull, who has a bachelor's degree in engineering physics, was working on making plastic objects from photopolymers at the company Ultra-Violet Products in California [4]. SLA uses an .stl file format to interpret the data in a CAD file, allowing these instructions to be communicated electronically to the 3D printer [4]. Along with shape, the instructions in the .stl file may also include information such as the color, texture, and thickness of the object to be printed [4].

Bioprinting. DOI: http://dx.doi.org/10.1016/B978-0-12-805369-0.00001-8

Hull later founded the company 3D Systems, which developed the first 3D printer, called a "SLA apparatus." [4] In 1988, 3D Systems introduced the first commercially available 3D printer, the SLA-250 [4]. Many other companies have since developed 3D printers for commercial applications, such as DTM Corporation, Z Corporation, Solidscape, and Objet Geometries [4]. Hull's work, as well as advances made by other researchers, has revolutionized manufacturing, and is poised to do the same in many other fields—including medicine [4].

Overview of Current Applications

Commercial Uses

Three-dimensional printing has been used by the manufacturing industry for decades, primarily to produce product prototypes [1,3]. Many manufacturers use large, fast 3D printers called "rapid prototyping machines" to create models and molds [5]. A large number of .stl files are available for commercial purposes [1]. Many of these printed objects are comparable to traditionally manufactured items [1].

Companies that use 3D printing for commercial medical applications have also emerged [6]. These include Helisys, Ultimateker, and Organovo, a company that uses 3D printing to fabricate living human tissue [6]. At present, however, the impact of 3D printing in medicine remains small [1]. Three-dimensional printing is currently a $700 million industry, with only $11 million (1.6%) invested in medical applications [1]. In the next 10 years, however, 3D printing is expected to grow into an $8.9 billion industry, with $1.9 billion (21%) projected to be spent on medical applications [1].

Consumer Uses

Three-dimensional printing technology is rapidly becoming easy and inexpensive enough to be used by consumers [3,5]. The accessibility of downloadable software from online repositories of 3D printing designs has proliferated, largely due to expanding applications and decreased cost [5–7]. It is now possible to print anything, from guns, clothing, and car parts to designer jewelry. Thousands of premade designs for 3D items are available for download, many of them for free.

Since 2006, two open-source 3D printers have become available to the public, Fab@Home (www.fabathome.org) and RepRap (www.reprap.org/wiki/RepRap) [3,4]. The availability of these open-source printers greatly lowered the barrier of entry for people who want to explore and develop new ideas for 3D printing [3]. These open-source systems allow anyone with a budget of about $1000 to build a 3D printer and start experimenting with new processes and materials [3].

This low-cost hardware and growing interest from hobbyists has spurred rapid growth in the consumer 3D printer market [5]. A relatively sophisticated 3D printer costs about $2500–$3000, and simpler models can be purchased for as little as $300–$400 [2,5]. For consumers who have difficulty printing 3D models themselves, several popular 3D printing services have emerged, such as Shapeways, (www.shapeways.com), Thingiverse (www.thingiverse.com), MyMiniFactory (www.myminifactory.com), and Threeding (www.threeding.com) [5].

4D Printing Market by Material (Programmable Carbon Fiber, Programmable Wood—Custom Printed Wood Grain, Programmable Textiles), End User (Aerospace, Automotive, Clothing, Construction, Defense, Healthcare & Utility) & Geography—Global Trends & Forecasts to 2019–25.

Four-dimensional printing is defined here as the technology in which the fourth dimension entails a change in form or function after the 3D printing of programmable material. In other words, 4D printing allows objects to be 3D printed and then to self-transform in shape and material property when exposed to a predetermined stimulus such as submersion in water, or exposure to heat, pressure, current, ultraviolet light, or some other source of energy.

Four-dimensional printing technology is expected to be commercialized in 2019. The global 4D printing market is expected to grow at a compound average growth rate (CAGR) of 42.95% between 2019 and 2025. The market is segmented on the basis of material segments into programmable carbon fiber, programmable wood grain, and programmable textiles. The programmable carbon fiber segment is expected to be the largest contributor to the overall market, with a share of ~62% of the market, in 2019.

This chapter describes the value chain for the 4D printing market, taking into consideration all the major stakeholders in the market and their role analysis. The report also provides a detailed study based on Porter's Five Forces framework for the market. All five major factors in these markets have been quantified using the internal key parameters governing each of them. We also discuss some of the leading players in the 3D printing industry along with their recent developments and other strategic business activities in this chapter. The competitive landscape section of this chapter outlines the potential of the key companies in the 3D printing industry.

Some of the key players in this market include 3D Systems Corporation (US), Autodesk, Inc. (US), Hewlett Packard Corp. (US), Stratasys Ltd. (US), ExOne Co. (US), Organovo Holdings, Inc. (US), Materialise NV (Belgium), and Dassault Systèmes SA (France).

- Healthcare industry
- Dental
- Medical

Medical applications for 3D printing are expanding rapidly and are expected to revolutionize health care [1]. Medical uses for 3D printing, both actual and potential, can be organized into several broad categories, including tissue and organ fabrication; creation of customized prosthetics, implants, and anatomical models; and pharmaceutical research regarding drug dosage forms, delivery, and discovery [6]. The application of 3D printing in medicine can provide many benefits, including the customization and personalization of medical products, drugs, and equipment; cost-effectiveness; increased productivity; the democratization of design and manufacturing; and enhanced collaboration [1,4,7–9].

- Utility

Biotechnology and bioengineering are both considered the parent disciplines of biomanufacturing while the remaining four fields are considered use areas. Any discoveries in the biotechnology or bioengineering fields will most likely trickle into areas of biomanufacturing. With that being said, bioengineering is the current key leader in recent new discoveries. Not commonly discussed is the role of bioinformatics in biomanufacturing. Data analysis from all of the biorelated disciplines must be examined for trends and irregularities from the hypotheses that were first initiated.

Bioprinting, defined as depositing cells, extracellular matrices, and other biologically relevant materials in user-defined patterns to build tissue constructs de novo or to build upon prefabricated scaffolds, is among one of the most promising techniques in tissue engineering.

Among the various technologies used for bioprinting, pressure-driven systems are most conducive to preserving cell viability. Herein, we explore the abilities of a novel bioprinter—Digilab, Inc.'s prototype cell printer—an automated liquid-handling device capable of delivering cell suspension in user-defined patterns onto standard cell culture substrates or custom-designed scaffolds. In this work, the feasibility of using the cell printer to deliver cell suspensions to biological sutures was explored.

Cell therapy using stem cells of various types shows promise to aid healing and regeneration in various ailments, including heart failure. Recent evidence suggests that delivering bone-marrow derived mesenchymal stem cells to the infarcted heart reduces infarct size and improves ventricular performance. Current cell delivery systems, however, have critical limitations such as inefficient cell retention, poor survival, and lack of targeted localization. Our laboratories have developed a method to produce discrete fibrin microthreads that can be bundled to form a suture and attached to a needle. These sutures can then be seeded with bone-marrow derived mesenchymal stem cells to deliver these cells to a precise location within the heart wall, both in terms of depth and surface localization. The efficiency of the process of seeding cells onto fibrin thread bundles (sutures) has previously been shown to be $11.8 \pm 3.9\%$, suggesting that 88% of the cells in suspension are not used. Considering that the proposed cell-therapy model for treatment of myocardial infarction contemplates use of autologous bone-marrow derived stem cells, an improvement in the efficiency of seeding cells onto the fibrin sutures is highly desirable.

The feasibility of using Digilab's prototype cell printer to deliver concentrated cell suspension containing human mesenchymal stem cells (hMSCs) directly onto a fibrin thread bundle was explored in this work, in order to determine if this technology could be adapted to seed cells onto such biological sutures. First, the effect of the printing process on the viability of hMSCs was assessed by comparing to cells dispensed manually using a handheld pipette. The viability of hMSCs 24 hours postdispensing using the cell printer was found to be $90.9 \pm 4.0\%$ and by manual pipetting was $90.6 \pm 8.2\%$ ($p = \text{ns}$). Thereafter a special bioreactor assembly composed of sterilizable Delrin plastic and stainless-steel pins was designed to mount fibrin thread bundles onto the deck of the cell printer, to deliver a suspension containing hMSCs on the bundles.

Highly targeted delivery of cell suspension directly onto fibrin thread bundles (average diameter 310 μm) was achieved with the bundle suspended in midair horizontally parallel to the printer's deck mounted on the bioreactor assembly. To compare seeding efficiency, fibrin thread bundles were simultaneously seeded with hMSCs using either the cell printer (or the current method tube-rotator method) and incubated for 24 hours. Seeded thread bundles were visualized using confocal microscopy and the number of cells per unit length of the bundle was determined for each group. The average seeding efficiency with the tube-rotator method was $7.0 \pm 0.03\%$ while the cell printer was $3.46 \pm 2.24\%$ ($p = \text{ns}$).

Ultimately, the cell printer was found to handle cells as gently as manual pipetting, preserving their viability, with the added abilities to dispense cells in user-defined patterns in an automated manner. With further development, such as localized temperature, gas and humidity control on the cell printer's deck to aid cell survival, the seeding efficiency is likely to improve. The feasibility of using this automated liquid-handling technology to deliver cells to biological scaffolds in specified patterns to develop vehicles for cell therapy is discussed in the following.

Seeding other cell types on other scaffolds along with selectively loading them with growth factors or multiple cell types can also be considered. In sum, the cell printer shows considerable potential to develop novel vehicles for cell therapy. It empowers researchers with a supervision- free, gentle, patterned cell-dispensing technique while preserving cell viability and a sterile environment. Looking forward, de novo biofabrication of tissue replicates on a small scale using the cell printer to dispense cells, extracellular matrices, and growth factors in different combinations is a very realistic possibility.

Examples of Technical Aspects of 3D Printing/Bioprinting

Bioprinting technology is based on the exact same principles of 3D printing, except in bioprinting the materials are safe for use with the human body.

I have personally owned three different 3D printers. In the following, I list the technical specifications of each.

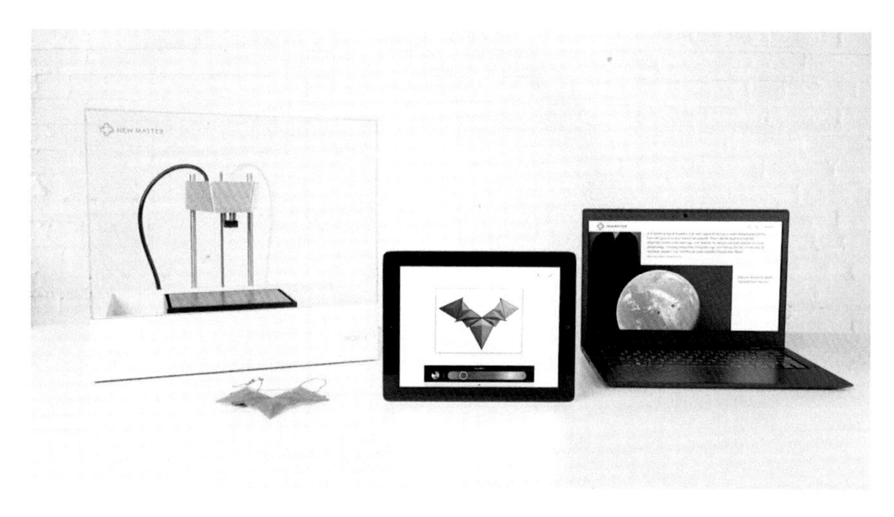

Figure 1.1 Complete setup ofMOD-t 3D printer from New Matter.

MOD-t From New Matter (@NewMatter): Cost $399 USD

MOD-t's key mechanical design breakthrough is their rack-and-pinion-driven XY bed. They've managed to do away with all the bearings, smooth rods, and threaded rod screws and replaced them with a simple mechanism of two pinions that the bed sits upon. The bed itself is a simple block of plastic with the racks attached underneath; this sits on the pinions and is then driven around the XY axes as they turn (Figs. 1.1–1.3).

Flashforge Dreamer: Cost $1299 USD

The Technique

Fused filament fabrication (FFF) is the most common method of 3D printing method that the Dreamer uses (Figs. 1.4 and 1.5). It works by melting plastic material called filament onto a print surface using high temperature. The filament solidifies after it cools down, which happens instantaneously after it is extruded from the print head. Three-dimensional objects are formed with the filament laying down multiple layers.

3D Printing Process

Three-dimensional printing involves three steps: 3D model design, slicing and exporting the 3D model, and making the print.

Print Specifications

Process	Fused Filament Fabrication
Build Material	Non-proprietary PLA filament, 1.75mm diameter
Maximum Object Size	150 x 100 x 125 mm (6 x 4 x 5 in)
Layer Resolution	Software selectable 0.2, 0.3, and 0.4 mm
Nozzle Diameter	0.4 mm

Dimensions & Weights (without spool)

Overall Dimensions	380 x 280 x 365 mm (15 x 11 x 14.5 in)
Product Weight	5 kg (11 lb)

Software & Electrical

Software	New Matter Desktop, Online Store, [Mobile1]
Supported File Types	STL, New Matter Store
Operating Systems	Windows 7+ [Macintosh OS X, Android, Apple iOS1]
Connectivity	Wi-Fi (802.11 b/g), USB 2.0
AC Power	100-240 VAC, 50/60 Hz, 80 W
Regulatory	CE, FCC Class B, CAN ICES-3B/NMB-3B

1 Planned future capabilities are enclosed within brackets. These capabilities are not planned for the initial release.

Figure 1.2 Technical specifications of the MOD-t 3D printer.

1. Designing the 3D Model: Currently, there are three ways of creating a 3D model.
   Designing From Scratch. One can use free CAD (computer-aided design) software such as Google Sketchup, Tinkercad, and Blender to design the 3D model. Or grab a copy of one of the commercial 3D design software packages such as AutoCAD, Maya, or Rhino3D.
   3D Scanners. An alternative method of creating a 3D model is to scan an object. Three-dimensional scanners work by digitizing a physical object, collecting its geometric data, and saving it to a file on the PC. There are also apps that can turn a mobile device into a 3D scanner.
   From the Cloud. The most popular way of obtaining a 3D model is to download it from websites that allow users to upload 3D models that they designed.

COMPARISON TABLE

	New Matter® MOD-t™	XYZ Printing da Vinci Jr. 1.0w	M3D Micro Retail Edition	XYZ Printing DaVinci 1.0	3D Systems Cube	MakerBot Replicator Mini
Price	$399	$399	$449	$499	$999	$1375
Wi-Fi Connectivity	Yes	Yes	No	No	No	No
Auto-Callibration	Yes	No	No	No	No	No
Browser-based Software	Yes	No	No	No	Yes	No
Max Print Dimensions	150 × 100 × 125 mm	150 × 150 × 150 mm	116 × 109 × 113 mm	200 × 200 × 200 mm	153 × 153 × 153 mm	140 × 140 × 135 mm
Print Material	PLA	PLA	PLA, ABS, Nylon	ABS	PLA, ABS	PLA, ABS
Layer Resolution	100 Microns	100 Microns	50 Microns	100 Microns	70 Microns	100 Microns
Filament	Non-proprietary	Proprietary	Non-proprietary	Proprietary	Proprietary	Non-proprietary
Max Print Speed	80 mm/s	90 mm/s	60 mm/s	150 mm/s	-	120 mm/s
Associated Storefront	Yes	Yes	No	Yes	Yes	Yes
Customer Support	Email, Phone, Chat, Blog, Forum, Web, Twitter, Facebook	Support Ticket, Phone, Email	Email Only	Support Ticket, Phone, Email	Phone Only	Chat (for registered products only), Phone, Support Case

Why do these product specifications matter?

- WiFi-connectivity combined with browser-based controls means you can use more devices with your 3D printer, like your smartphone or tablet.
- Some 3D printers require a labor-intensive calibration process. Auto-calibration takes 5 to 10 minutes and doesn't require any additional effort or attention.
- PLA filament is recommended over ABS filament for homes and classrooms as it does not produce a strong odor or require special ventilation.
- Layer resolution affects smoothness of 3D prints. With smaller layer resolution, striations on the sides of 3D prints will be smaller.
- 3D Printers using proprietary filament only accept filament produced by the 3D printer manufacturer and can be more expensive, despite using the same material. Printers that use non-proprietary filament allow use of filament from a variety of manufacturers.
- 3D printers with associated storefronts have a design library for you to select 3D designs from. All of these 3D printers also allow users to 3D print user-created designs or designs downloaded from the internet.

Figure 1.3 FlashForge Dreamer technical specifications.

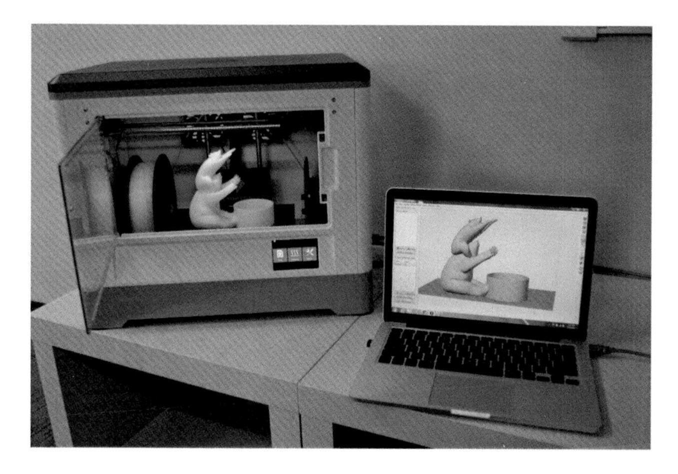

Figure 1.4 FlashForge Dreamer complete setup.

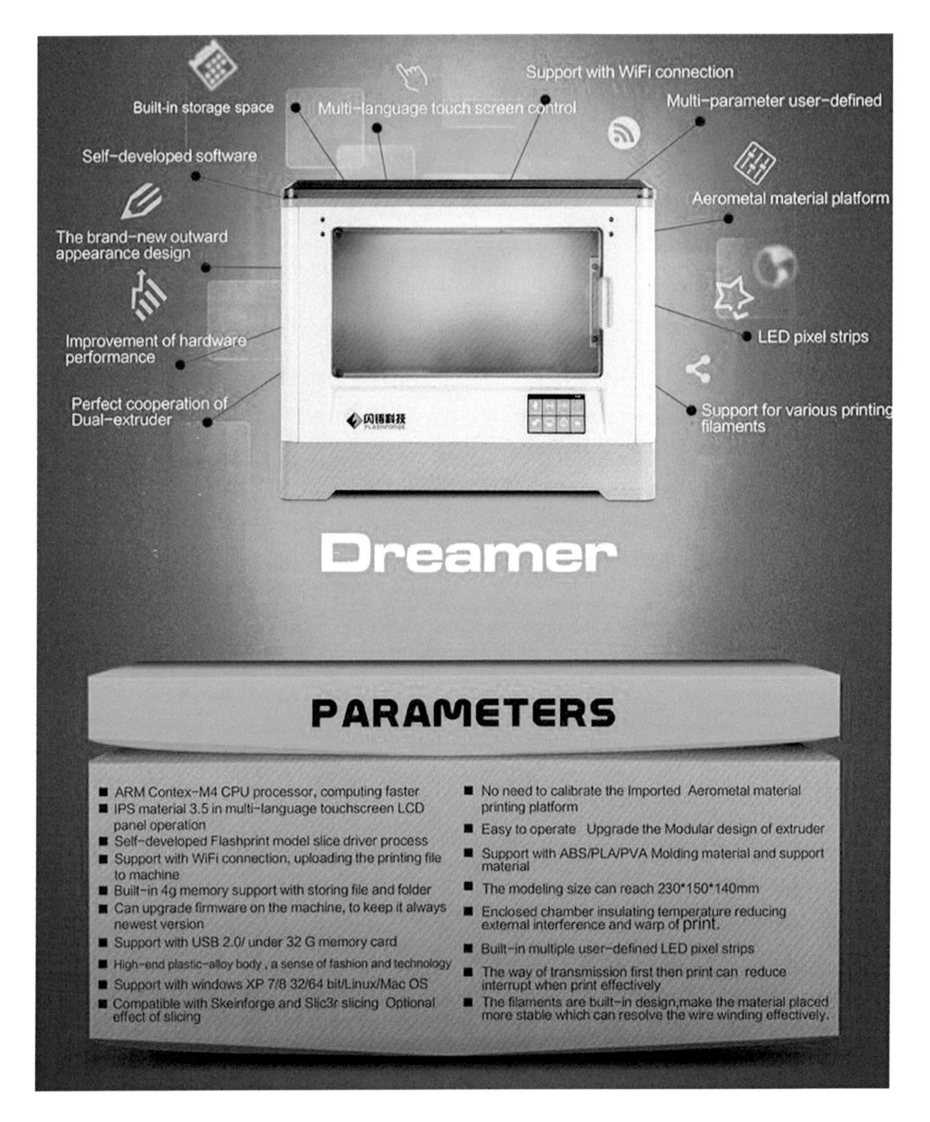

Figure 1.5 FlashForge Dreamer technical specifications.

2. Slicing and Exporting the 3D Model: A slicing software is required to process and interpret the 3D model into the language FFF 3D printers understand. The FlashPrint is the slicing software used for the FlashForge Dreamer.
   FlashPrint will slice the 3D model into numerous layers and output it as a .g file, which is the format read by Dreamer.

The file is then transferred to Dreamer by USB cable, SD card, or WiFi.

3. Making the Print: Once the output file has been transferred to Dreamer, it will start to turn the 3D model into a physical object by laying down layers of filament.

iBox Nano: Cost $400–$700 USD

The iBox Nano works over WiFi, so one can print without being tethered to the 3D resin printer.

Many of the resin printers on the market use Digital Light Processing (DLP) technology to create and control the UV light used to cure the resin. Using a DLP projector introduces a few issues such as low bulb life and cooling fan noise. DLP projector bulbs will need to be replaced at their service intervals, generally at 2000–8000 hours of use. Even before they fail they will suffer a noticeable loss in output power causing reduced print quality, or no print at all. These bulbs can cost hundreds of dollars, and all of them will need to be replaced.

DLP projectors and laser-based SLA resin printers require cooling fans to be running 100% of the time. The DLPs use them to cool the bulbs to extend their life, and the laser systems use them to cool the galvanometer drivers to extend their life. This printer uses very little power and generates almost no heat, thus not requiring a cooling fan and its associated noise (Figs. 1.6–1.9).

Most consumers buy a large 3D printer and statistically only print small items. The reasons behind this are material cost, print time, and the nature of the items that are typically 3D printed. Large build plates on resin printers can also be a disadvantage. Unlike fused deposition modeling (FDM) (filament) printers the resin in the build tray is inadvertently exposed to UV radiation from sources such as indoor lighting. Over time, the resin in the vat degrades, which leads to failed prints. Resin is a consumable, so when printing small things, the resin printer will save money.

The iBox Nano is designed for the home user who wants to print small-to-average-sized 3D objects with good resolution without having a large noisy printer intruding on their workspace. It is small, quiet, inexpensive, and portable.

Figure 1.6 iBox Nano 3D resin printer.

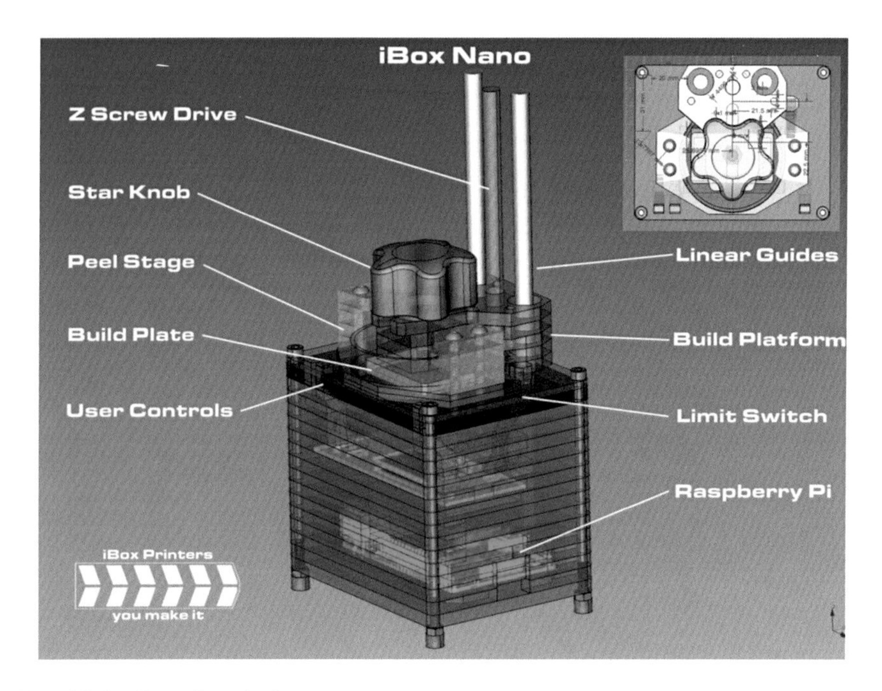

Figure 1.7 iBox Nano schematic of parts.

	Price	Shipping	Resolution	Technology
iBox Nano	$299	$25	328 Microns	LCD - Resin
Form 1	$3299	$68	300 Microns	UV-Laser - Resin
B9 Creator	$5495	$100		Projector - Resin
Titan 1	$2899	~$80-100	100 Microns	Projector - Resin
Pirate3D	$699		100 Microns	FDM (filament)
Replicator 2x	$2799		100 Microns	FDM
Pegasus	$3499	$75	250 Microns	Laser SLA
LittleRP	$1250			
Robox	$1300	~$50	300 Microns	Dual FDM
Zeus	$2499	$69	125 Microns	FDM

Figure 1.8 The iBox Nano has almost three times the z *resolution of the next most expensive 3D printer.*

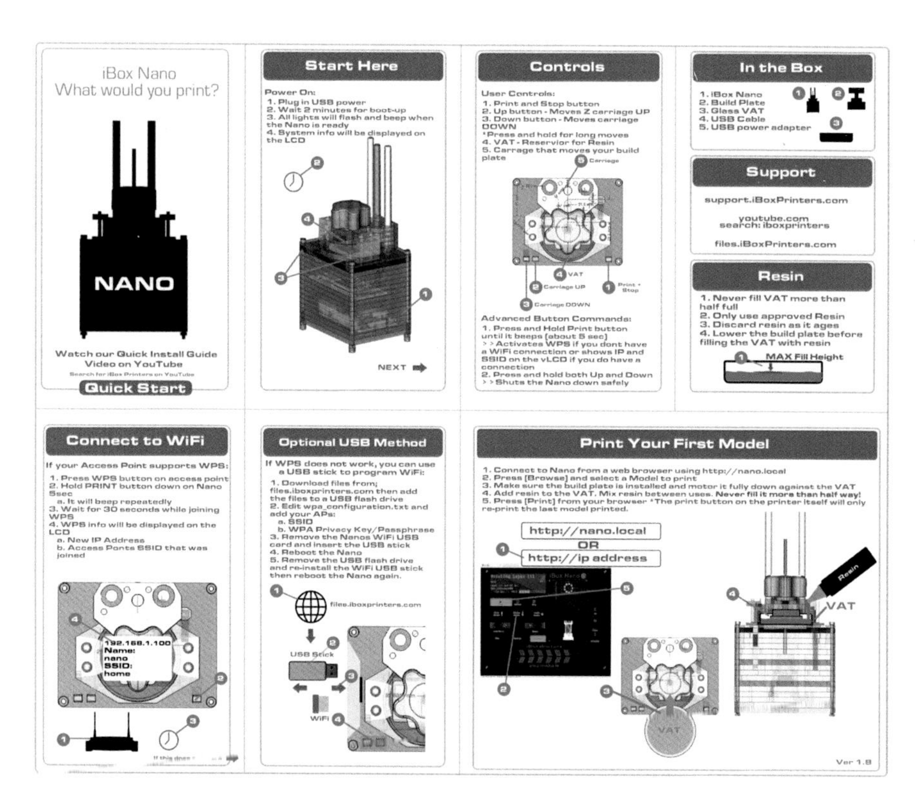

Figure 1.9 iBox Nano user guide.

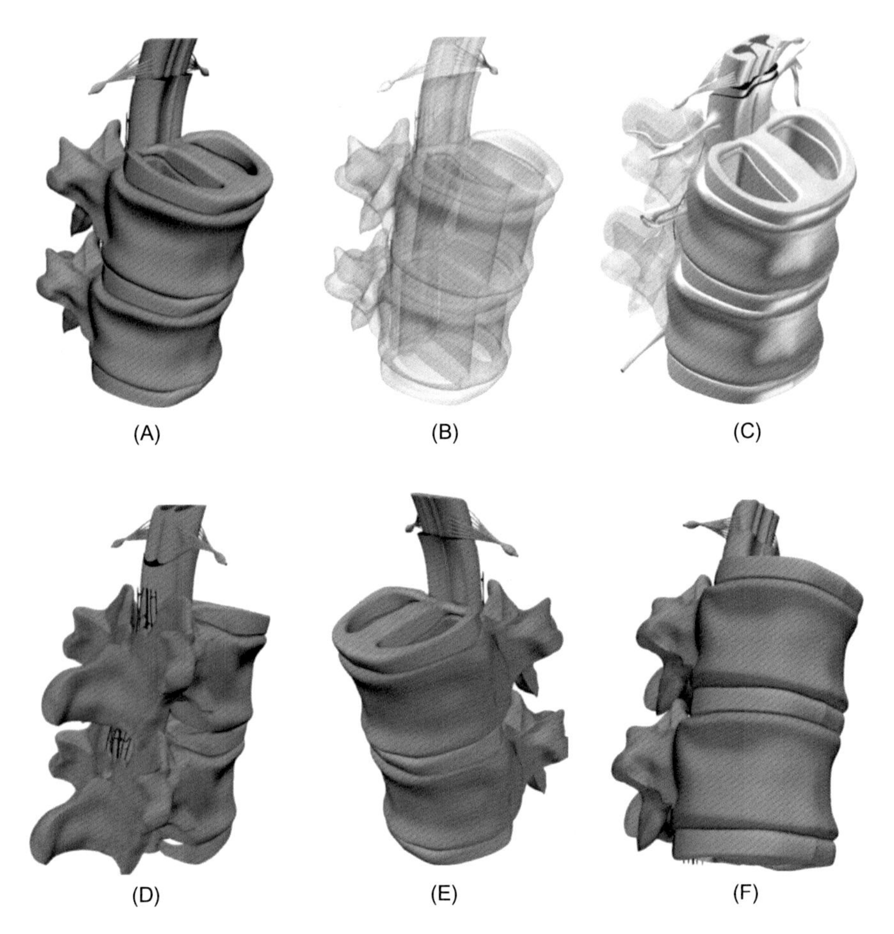

Figure 1.10 (A–F) Vertebrae STEP CAD file. A 3D printer uses instructions in a digital file to create a physical object [12].

There are many 3D printing processes, and they use varying printer technologies, speeds, and resolutions, and even more types of exotic materials [3]. These technologies can build a 3D object in almost any shape imaginable as defined in a CAD file (Fig. 1.10A–E) [3]. In a basic setup, the 3D printer first follows the instructions in the CAD file to build the foundation for the object, moving the printhead along the x–y plane [9,10]. The printer then continues to follow the instructions, moving the printhead along the z-axis to build the object vertically layer by layer [9]. It is important to note that 2D radiographic images, such as x-rays, magnetic resonance imaging, or

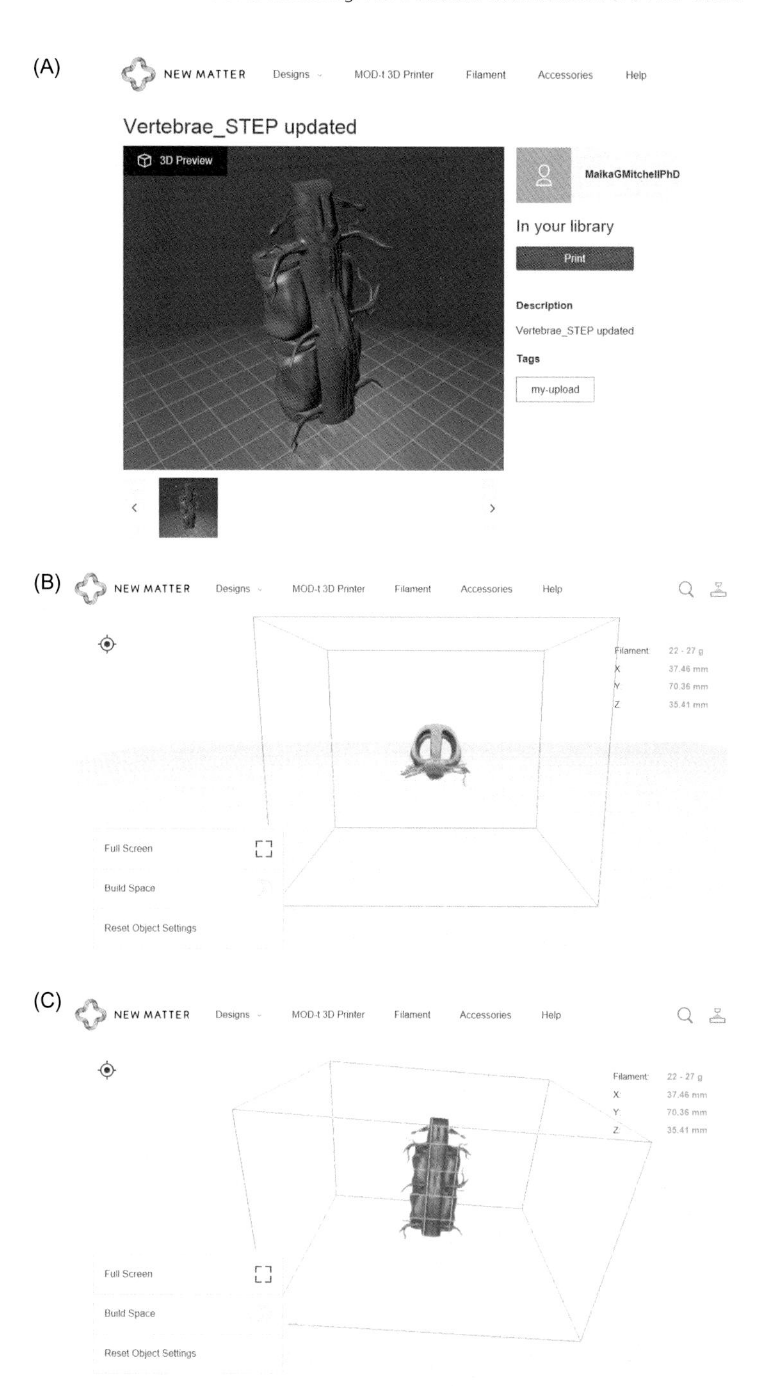

Figure 1.11 (A) Vertebrae STEP Meshmixer file on the Mod-t New Matter software; (B) Front view of vertebrae STEP CAD file; (A, B, & C) Vertebrae STEP CAD file; ((A) Front view and (B) Bottom view).

Figure 1.12 (A) iBox Nano user interface and software on Netbook PC with vertebrae STEP CAD svg file; (B) iBox Nano vertebrae STEP 3D resin print; (C) iBox Nano vertebrae STEP 3D resin print (front view); (D) Vertebrae STEP printed in blue resin on the iBox Nano 3D printer in different rotational views.

computerized tomography scans, can be converted to digital 3D print files, allowing the creation of complex, customized anatomical and medical structures (Figs. 1.11–1.13) [8,9,11].

Radiographic images can be converted to 3D print files to create complex, customized anatomical and medical structures [12–19].

(A)

(B)

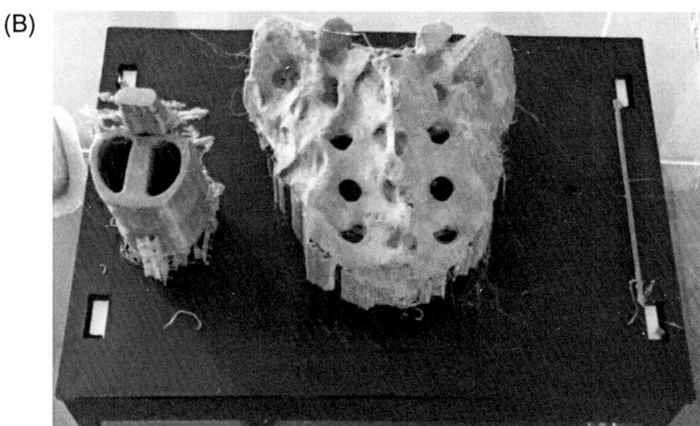

Figure 1.13 (A) Vertebrae STEP and sacrum printed with bronze and plastic composite polylactic acid (PLA) on the MOD-t by New Matter 3D printer (front view); (B) Vertebrae STEP and sacrum printed with bronze and plastic composite PLA on the MOD-t by New Matter 3D printer (top view).

REFERENCES

[1] Schubert C, van Langeveld MC, Donoso LA. Innovations in 3D printing: a 3D overview from optics to organs. Br J Ophthalmol 2014;98(2):159–61.

[2] Science and society: experts warn against bans on 3D printing. Science. 2013;342(6157):439.

[3] Lipson H. New world of 3-D printing offers "completely new ways of thinking:" Q&A with author, engineer, and 3-D printing expert Hod Lipson. IEEE Pulse 2013;4(6):12–14.

[4] Gross BC, Erkal JL, Lockwood SY, et al. Evaluation of 3D printing and its potential impact on biotechnology and the chemical sciences. Anal Chem 2014;86(7):3240–53.

[5] Hoy MB. 3D printing: making things at the library. Med Ref Serv Q 2013;32(1):94–9.

[6] Klein GT, Lu Y, Wang MY. 3D printing and neurosurgery—ready for prime time? World Neurosurg 2013;80(3–4):233–5.

[7] Mertz L. Dream it, design it, print it in 3-D: What can 3-D printing do for you?. IEEE Pulse 2013;4(6):15–21.

[8] Banks J. Adding value in additive manufacturing: researchers in the United Kingdom and Europe look to 3D printing for customization. IEEE Pulse 2013;4(6):22–6.

[9] Ursan I, Chiu L, Pierce A. Three-dimensional drug printing: a structured review. J Am Pharm Assoc 2013;53(2):136–44.

[10] Bartlett S. Printing organs on demand. Lancet Respir Med 2013;1(9):684.

[11] Cui X, Boland T, D'Lima DD, Lotz MK. Thermal inkjet printing in tissue engineering and regenerative medicine. Recent Pat Drug Deliv Formul 2012;6(2):149–55.

[12] 3D Print Exchange. National Institutes of Health; Available at: http://3dprint.nih.gov. Accessed August 9, 2016.

[13] Ozbolat IT, Yu Y. Bioprinting toward organ fabrication: challenges and future trends. IEEE Trans Biomed Eng 2013;60(3):691–9.

[14] Bertassoni L, Cecconi M, Manoharan V, et al. Hydrogel bioprinted microchannel networks for vascularization of tissue engineering constructs. Lab on a Chip 2014;14(13):2202.

[15] Centers for Disease Control and Prevention Colorectal cancer statistics. Sep 2, 2014. Available at: http://www.cdc.gov/cancer/colorectal/statistics. Accessed September 17, 2014.

[16] Khaled SA, Burley JC, Alexander MR, Roberts CJ. Desktop 3D printing of controlled release pharmaceutical bilayer tablets. Int J Pharm 2014;461(1–2):105–11.

[17] Plastics Today. FDA tackles opportunities, challenges, of 3D printed medical devices. Jun 2, 2014. Available at: http://www.plasticstoday.com/articles/FDA-tackles-opportunities-challenges-3D-printed-medical-devices-140602. Accessed July 9, 2014.

[18] Food and Drug Administration Public workshop—additive manufacturing of medical devices: an interactive discussion on the technical considerations of 3D printing. Sep 3, 2014. Available at: http://www.fda.gov/medicaldevices/newsevents/workshopsconferences/ucm397324.htm. Accessed September 17, 2014.

[19] Cheow WS, Kiew TY, Hadinoto K. Combining inkjet printing and amorphous nanonization to prepare personalized dosage forms of poorly-soluble drugs. Eur J Pharm Biopharm 2015;96:314–21.

ONLINE RESOURCES ON BIOMANUFACTURING

BayBio http://www.baybio.org/wt/page/index

Biotechnology Industry Organization (BIO) http://www.bio.org/

BioBasics: Industrial Biotechnology http://biobasics.gc.ca/english/View.asp?x=614

CheckBiotech http://www.checkbiotech.org/ Genencor International http://www.genencor.com/cms/connect/genencor Genentech http://www.gene.com/

Genetic Engineering & Biotechnology News http://www.geneng-news.com/

Northeast Biomanufacturing Center & Collaborative http://www.biomanufacturing.org/index.htm

Novartis http://www.novartis.com/

Novozymes http://www.novozymes.com/en

Science Daily http://www.sciencedaily.com/news/plants_animals/biotechnology/

CHAPTER 2

Reproducing Cells Is Nothing New—A Historical Prospective

For years, scientists have been growing cells in laboratories, including skin tissue, blood vessels, and other cell cultures from various organs. Replicating and growing cells in petri dishes is nothing new, and the science surrounding this is constantly advancing. However, 3D printing offers an opportunity to print an entire organ, not just pieces of one, which has the potential to drastically reduce the cost of these processes because of the cells and other materials used.

Three-dimensional bioprinting began with the cell culture of the well-noted HeLa cell line. A HeLa cell is a cell type in an immortal cell line used in scientific research, and is the oldest and most commonly used human cell line [1]. The line was derived from cervical cancer cells taken in early 1951 [2] from Henrietta Lacks, a patient who died of her cancer on October 4, 1951. The cell line was found to be remarkably durable and prolific, which led to it contaminating many other cell lines used in research [3,4].

The cells from Lacks' tumor were taken without her knowledge or consent by researcher George Gey, who found that they could be kept alive [5]. Before this, cells cultured from other cells would only survive for a few days. Scientists spent more time trying to keep the cells alive than performing actual research on them, but some cells from Lacks' tumor sample behaved differently than others. George Gey was able to isolate one specific cell, multiply it, and start a cell line. Gey named the sample HeLa, after the initial letters of Henrietta Lacks' name. They were the first human cells grown in a lab that were "immortal," i.e., they did not die after a few cell divisions, and could be used to conduct many experiments. This represented an enormous boon to medical and biological research [4].

Bioprinting. DOI: http://dx.doi.org/10.1016/B978-0-12-805369-0.00002-X

The stable growth of HeLa enabled a researcher at the University of Minnesota hospital to successfully grow polio virus, enabling the development of a vaccine [6], and by 1954, a vaccine was developed for polio using these cells by Jonas Salk [4,7]. To test Salk's new vaccine, the cells were put into mass production in the first-ever cell production factory [8].

In 1955, HeLa cells were the first human cells successfully cloned [9].

Demand for the HeLa cells quickly grew. Since they were put into mass production, Lacks' cells have been used by scientists around the globe for "research into cancer, AIDS, the effects of radiation and toxic substances, gene mapping, and countless other scientific pursuits" [7]. HeLa cells have been used to test human sensitivity to tape, glue, cosmetics, and many other products [4]. Scientists have grown some 20 tons of her cells [4,10], and there are almost 11,000 patents involving HeLa cells [4].

The cells were propagated by George Otto Gey shortly before Lacks died of her cancer in 1951. This was the first human cell line to prove successful in vitro, which was a scientific achievement with profound future benefit to medical research. Gey freely donated these cells along with the tools and processes that his lab developed to any scientist requesting them simply for the benefit of science. Neither Lacks nor her family gave permission to harvest the cells but, at that time, permission was neither required nor customarily sought [11]. The cells were later commercialized, although never patented in their original form. There was no requirement at that time (or at the present) to inform patients or their relatives about such matters because discarded material or material obtained during surgery, diagnosis, or therapy was the property of the physician or the medical institution (this currently requires ethical approval and patient consent in the United Kingdom). This issue and Lacks' situation were brought up in the Supreme Court of California case of Moore v. Regents of the University of California. The court ruled that a person's discarded tissue and cells are not his or her property and can be commercialized [12].

At first, the cell line was said to be named after a "Helen Lane" or "Helen Larson," in order to preserve Lacks' anonymity. Despite this attempt, her real name was used by the press within a few years of her death. These cells are treated as cancer cells, as they are descended from a biopsy taken from a visible lesion on the cervix as part of Lacks' diagnosis of cancer.

HeLa cells, like other cell lines, are termed "immortal" in that they can divide an unlimited number of times in a laboratory cell culture plate as long as fundamental cell survival conditions are met (i.e., being maintained and sustained in a suitable environment). There are many strains of HeLa cells as they continue to mutate in cell cultures, but all HeLa cells are descended from the same tumor cells removed from Lacks. The total number of HeLa cells that have been propagated in cell culture far exceeds the total number of cells that were in Henrietta Lacks' body [13].

USE IN RESEARCH

HeLa cells were used by Jonas Salk to test the first polio vaccine in the 1950s. They were observed to be easily infected by poliomyelitis, causing infected cells to die [2]. This made HeLa cells highly desirable for polio vaccine testing since results could be easily obtained. A large volume of HeLa cells were needed for the testing of Salk's polio vaccine, prompting the National Foundation for Infantile Paralysis (NFIP) to find a facility capable of mass-producing HeLa cells [14]. In the spring of 1953, a cell culture factory was established at Tuskegee University to supply Salk and other labs with HeLa cells [15]. Less than a year later, Salk's vaccine was ready for human trials [16].

HeLa cells were also the first human cells to be successfully cloned in 1955 by Theodore Puck and Philip I Marcus at the University of Colorado, Denver [9]. Since that time, HeLa cells have been used for "research into cancer, AIDS, the effects of radiation and toxic substances, gene mapping, and many other scientific pursuits" [7]. According to author Rebecca Skloot, by 2009, "more than 60,000 scientific articles had been published about research done on HeLa, and that number was increasing steadily at a rate of more than 300 papers each month" [12].

HeLa cells have been used in testing how parvo virus infects cells of humans, HeLa, dogs, and cats [17]. These cells have also been used to study viruses such as the Oropouche virus (OROV), which causes the disruption of cells in culture, where cells begin to degenerate shortly after they are infected, causing viral induction of apoptosis [18]. HeLa cells have been used to study the expression of the papillomavirus E2 and apoptosis [19]. HeLa cells have also been used to study canine distemper virus' ability to induce apoptosis in cancer cell lines [20], which could play an important role in developing treatments for tumor cells resistant to radiation and chemotherapy [20]. HeLa cells have also been used in a number of cancer studies, including those involving sex steroid hormones such as estradiol, estrogen, and estrogen receptors, along with estrogen-like compounds such as quercetin and its cancer-reducing properties [21]. There have also been studies on HeLa cells and the effects of flavonoids and antioxidants with estradiol on cancer cell proliferation. HeLa cells were used to investigate the phytochemical compounds and the fundamental mechanism of the anticancer activity of the ethanolic extract of mango peel (EEMP), which was found to contain various phenolic compounds and to activate death of human cervical malignant HeLa cells through apoptosis, which suggests that EEMP may help to prevent cervical cancer as well as other types of cancers [22].

In 2011, HeLa cells were used to test novel heptamethine dyes such as IR-808 and other analogs currently being explored for their unique use in medical diagnostics, the development of theranostics, the individualized treatment of cancer patients with the aid of PDT, coadministration with other drugs, and irradiation [23,24]. HeLa cells have been used in research involving fullerenes to induce apoptosis as a part of photodynamic therapy, as well as in in vitro cancer research using cell lines [25]. Further, HeLa cells have also been used to define cancer markers in RNA, and have been used to establish an RNAi-based identification system and interference of specific cancer cells [26].

2D AND 3D CELL CULTURE APPLICATIONS LEADING TO BIOPRINTING

Characteristics of 3D Cell Cultures Versus Traditional 2D Cell Cultures

Fig. 2.1 shows the schematic diagrams of traditional 2D cell cultures and three typical 3D cell cultures. While the traditional 2D culture usually grows cells into a monolayer on glass or, more commonly, tissue culture polystyrene plastic flasks (Fig. 2.1A) 3D cell cultures grow cells into 3D aggregates/spheroids using a scaffold/matrix (Fig. 2.1B and C) or in a scaffold-free manner (Fig. 2.1D). Scaffold/matrix-based 3D cultures can

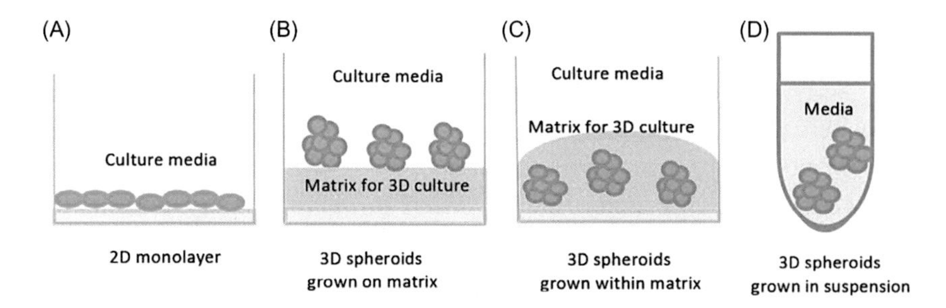

Figure 2.1 Schematic diagrams of traditional two-dimensional (2D) monolayer cell culture (A) and three typical three-dimensional (3D) cell culture systems: cell spheroids/aggregates grown on matrix (B), cells embedded within matrix (C), or (D) scaffold-free cell.

be generated by seeding cells on an acellular 3D matrix or by dispersing cells in a liquid matrix followed by solidification or polymerization. Commonly used scaffold/matrix materials include biologically derived scaffold systems and synthetic-based materials. Commercially available products such as BD Matrigel basement membrane matrix (BD Sciences), Cultrex basement membrane extract (BME; Trevigen), and hyaluronic acid are commonly used biologically derived matrixes. Polyethylene glycol (PEG), polyvinyl alcohol (PVA), polylactide-*co*-glycolide (PLG), and polycaprolactone (PLA) are common materials used to form synthetic scaffolds [5,9,16,17]. Scaffold-free 3D cell spheroids can be generated in suspensions by the forced floating method, the hanging drop method, or agitation-based approaches [5]. More detailed information about a number of commercially available 3D cell culture systems, their specific features, as well as advantages and disadvantages of each type of the above-mentioned methods can be found in two of the most recent reviews [5,17]. With each of these methods, cells grow naturally in a 3D environment, allowing cells to interact with each other, the extracellular matrix (ECM), and their microenvironment. In turn, these interactions in such a 3D spatial arrangement affect a range of cellular functions, including cell proliferation, differentiation, morphology, gene and protein expression, and cellular responses to external stimuli. The following section reviews the characteristics of cells in 3D cultures in comparison to cells in traditional 2D culture.

Growth Conditions, Cell Morphology, and Populations in 2D and 3D Cultures

In traditional 2D monolayer culture, cells adhere and grow on a flat surface. Such a monolayer setting allows all of the cells to receive a homogenous amount of nutrients and growth factors from the medium during growth [18]. The monolayer is mainly composed of proliferating cells, since necrotic cells are usually detached from the surfaces and easily removed during medium change. Cells grown in 2D culture are usually more flat and stretched than they would appear in vivo. The abnormal cell morphology in 2D culture influences many cellular processes including cell proliferation, differentiation, apoptosis, and gene and protein expression [9]. As a result, 2D-cultured cells may not behave as they would in the body because this model does not adequately mimic the in vivo microenvironment [19]. Technologies such as nano-patterning, which mimics the topographical features of the ECM, have been investigated to improve cellular function and behavior in 2D cell culture [20,21]. However, whether or not these changes in cell function better emulate in vivo behaviors is still under investigation. The traditional 2D cell culture is still the most common in vitro test platform in drug screening.

As opposed to 2D monolayer culture, when grown in 3D culture systems, cells form aggregates or spheroids within a matrix, on a matrix, or in a suspension medium. In cell aggregates/spheroids, cell–cell interactions and cell–ECM interactions more closely mimic the natural environment found in vivo, so that the cell morphology closely resembles its natural shape in the body. In addition, 3D spheroids are comprised of cells in various stages, usually including proliferating, quiescent, apoptotic, hypoxic, and necrotic cells [22,23]. The outer layers of a spheroid, which is highly exposed to the medium, are mainly comprised of viable, proliferating cells [23]. The core cells receive less oxygen, growth factors, and nutrients from the medium, and tend to be in a quiescent or hypoxic state. Such cellular heterogeneity is very similar to in vivo tissues, particularly in tumors. Since the morphology and the interactions of cells grown in 3D culture is more similar to what occurs in vivo, the cellular processes of these cells also closely emulate what is seen in vivo.

The proliferation rates of cells cultured in 3D and 2D are usually different, and are cell line and matrix dependent. A variety of cell lines showed reduced proliferation rate in 3D cultures compared to those cultured in 2D [25–29].

REFERENCES

[1] Rahbari R, Sheahan T, Modes V, Collier P, Macfarlane C, Badge RM. A novel L1 retrotransposon marker for HeLa cell line identification. BioTechniques 2009;46(4):277–84.

[2] Scherer WF, Syverton JT, Gey GO. Studies on the propagation in vitro of poliomyelitis viruses: IV. Viral multiplication in a stable strain of human malignant epithelial cells (strain HeLa) derived from an epidermoid carcinoma of the cervix. J Exp Med 1953;97 (5):695–710. Available from: http://dx.doi.org/10.1084/jem.97.5.695.

[3] Capes-Davis A, Theodosopoulos G, Atkin I, Drexler HG, Kohara A, MacLeod RA, et al. Check your cultures! A list of cross-contaminated or misidentified cell lines. Int J Cancer 2010;127(1):1–8.

[4] Batts DW. Cancer cells killed Henrietta Lacks – then made her immortal. The Virginian-Pilot, pp. 1, 12–14; 2010-05-10. Retrieved 2016-03-17.

[5] Claiborne R, Wright IV S, How one woman's cells changed medicine. ABC World News; 2010-01-31. Retrieved 2016-08-19.

[6] Scherer WF, Syverton JT, Gey GO. Studies on the propagation in vitro of poliomyelitis viruses. J Exp Med May 1, 1953;97(5):695–710.

[7] Smith V. Wonder woman: the life, death, and life after death of Henrietta Lacks, unwitting heroine of modern medical science. Baltimore City Paper; 2002-04-17. Retrieved 2016-08-19.

[8] Skloot Rebecca. The Immortal Life of Henrietta Lacks. New York: Random House; 2010, p. 96.

[9] Puck TT, Marcus PI. A rapid method for viable cell titration and clone production with HeLa cells in tissue culture: the use of X-irradiated cells to supply conditioning factors. Proc Natl Acad Sci U S A 1955;41(7):432–7. Bibcode:1955PNAS...41..432P.

[10] Margonelli L. Eternal Life. New York Times, New York; February 5, 2010. Retrieved 23 April 2016.

[11] Washington H. Henrietta Lacks: an unsung hero. Emerge Magazine; October 1994.

[12] Skloot R. The immortal life of Henrietta Lacks. New York: Crown/Random House; 2010. ISBN 978-1-4000-5217-2.

[13] Sharrer T. "HeLa" herself. The Scientist 2006;20(7):22.

[14] Masters JR. TIMELINEHeLa cells 50 years on: the good, the bad and the ugly. Nat Rev Cancer 2002;2(4):315–19.

[15] Turner T. Development of the polio vaccine: a historical perspective of Tuskegee University's role in mass production and distribution of HeLa cells. J Health Care Poor Underserved 2012;23(4a):5–10.

[16] Brownlee KA. Statistics of the 1954 polio vaccine trials*. J Am Stat Assoc 1955;50 (272):1005–13.

[17] Parker J, Murphy W, Wang D, O'Brien S, Parrish C. Canine and feline parvoviruses can use human or feline transferrin receptors to bind, enter, and infect cells. J Virol 2001;75 (8):3896–902. Available from: http://dx.doi.org/10.1128/JVI.75.8.3896-3902.2001.

[18] Acrani GO, Gomes R, Proença-Módena JL, da Silva AF, Carminati PO, Silva ML, et al. Apoptosis induced by Oropouche virus infection in HeLa cells is dependent on virus protein

expression. Virus Res 2010;149(1):56–63. Available from: http://dx.doi.org/10.1016/j.virusres.2009.12.013.

[19] Hou SY, Wu S, Chiang C. Transcriptional activity among high and low risk human papillomavirus E2 proteins correlates with E2 DNA binding. J Biol Chem 2002;277(47):45619–29. Available from: http://dx.doi.org/10.1074/jbc.M206829200.

[20] Del Puerto HL, Martins AS, Milsted A, Souza-Fagundes EM, Braz GF, Hissa B, et al. Canine distemper virus induces apoptosis in cervical tumor derived cell lines. Virol J 2011;8 (1):334.

[21] Bulzomi P, Galluzzo P, Bolli A, Leone S, Acconcia F, Marino M. The pro-apoptotic effect of quercetin in cancer cell lines requires ERβ-dependent signals. J Cell Physiol 2012;227 (5):1891–8.

[22] Kim H, Kim H, Mosaddik A, Gyawali R, Ahn KS, Cho SK. Induction of apoptosis by ethanolic extract of mango peel and comparative analysis of the chemical consists of mango peel and flesh. Food Chem 2012;133(2):416–22.

[23] Tan X, Luo S, Wang D, Su Y, Cheng T, Shi C. A NIR heptamethine Dye with intrinsic cancer targeting, imaging and photosynthesizing properties. J Biomater China 2011;33 (7):2230–9.

[24] Pene F, Courtine E, Cariou A, Mira JP. Toward theranostics. Crit Care Med 2009;37: S50–8.

[25] Briiuner T, Hulser DF. Tumor Cell Invasion and Gap Junctional Communication. Invasion Metastasis 1990;10:31–4.

[26] Xie Z, Wroblewska L, Prochazka L, Weiss R, Benenson Y. Multi-input RNAi-based logic circuit for identification of specific cancer cells. Science 2011;333(6047):1307–11. Bibcode:2011Sci...333.1307X

[27] Edmondson R, et al. Three-dimensional cell culture systems and their applications in drug discovery and cell-based biosensors. Assay Drug Dev Technol 2014;12.4:207–18.

[28] Birgersdotter A, Sandberg R, Ernberg I. Gene expression perturbation in vitro—a growing case for three-dimensional (3D) culture systems. Semin Cancer Biol 2005;15:405–12.

[29] Weaver VM, Petersen OW, Wang F, et al. Reversion of the malignant phenotype of human breast cells in three-dimensional culture and in vivo by integrin blocking antibodies. J Cell Biol 1997;137:231–45.

CHAPTER 3

Bioprinting Versus 3D Printing

One of the first experiments using bioprinting involved creating liver tissue. Organovo used spheroids of parenchymal (or fundamental) liver cells loaded into a syringe. In another syringe, nonparenchymal liver cells and hydrogel, which fuse together to create a bioink, is loaded. The bioink makes a mold in the cell dish, and the liver cells fill up the rest of the dish. When the cells are put in an incubator, they fuse together even more to form the full liver tissue.

This type of 3D printing aims to allow scientists and medical researchers to build an organ, layer by layer, using scanners and printers traditionally reserved for auto design, model building, and product prototyping (Fig. 3.1).

3D PRINTERS (AND TYPES) I PERSONALLY OWN AND USE

Micro3D (Cost in June 2015: USD $250)

Technical Specifications

Supports many different materials: ABS, PLA, nylon, professional, chameleon

50–350 μm layer resolution

15 μm *X* and *Y* positioning accuracy

Filament: standard 1.75 mm. 1/2 lb rolls fit within print bed and allow you to try a variety of materials and colors for less! Standard filament rolls also supported.

Print height: 116 mm (4.6″)

Base print area: 109 mm × 113 mm

Print area above: 74 mm: 91 mm × 84 mm

Removable print bed size: 128 mm × 128 mm

Bioprinting. DOI: http://dx.doi.org/10.1016/B978-0-12-805369-0.00003-1

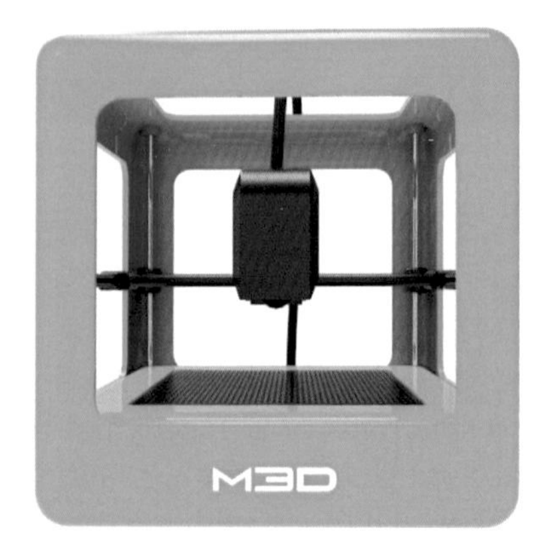

Figure 3.1 M3D SLA 3D printer photos in use with an HP PC Netbook.

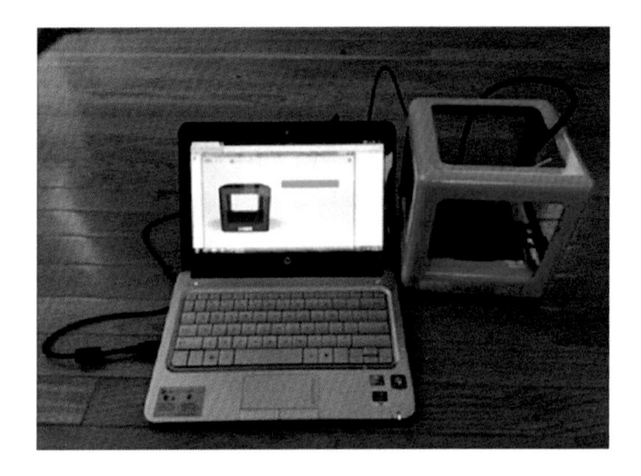

Figure 3.2 M3D SLA 3D printer photos in use with an HP PC Netbook.

Printer dimensions: It's a cube, 7.3 in (185 mm) per side.

Printer weight: 1 kg (2.2 lbs)

Package weight: 2 kg–2.7 kg (4.4 lbs–6 lbs)

USB compatible

SOFTWARE

M3D software for an effortless, plug-and-play experience

File types supported: .stl, .obj

Compatible with Mac and PC (Fig. 3.2).

The Micro3D was the first 3D printer I bought. The cost is perfect for someone just starting out and getting to know the technology. I hooked it up to my HP Netbook and was up and running within 30 minutes (Fig. 3.3).

I used an imaging scan of a knee tumor prosthetic and turned it into a reality a few hours later (Figs. 3.3 and 3.4).

Duct tape was used as the print kept sliding off of the print bed (see Figs. 3.5–3.8).

Figure 3.3 M3D SLA 3D printer photos in use with an HP PC Netbook.

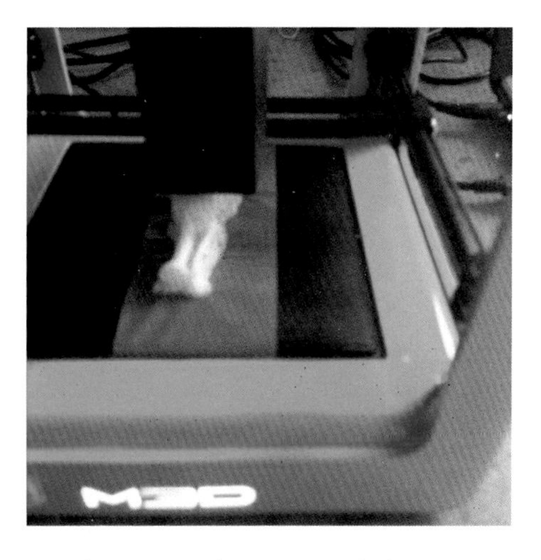

Figure 3.4 M3D SLA 3D printer photos in use with an HP PC Netbook.

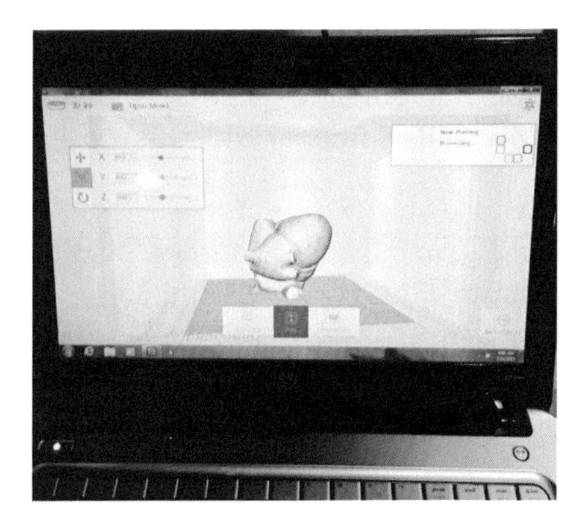

Figure 3.5 M3D SLA 3D printer photos in use with an HP PC Netbook.

Figure 3.6 M3D SLA 3D printer photos in use with an HP PC Netbook.

Human Heart File From Thingiverse

All 3D print files were creating using CT/MRI/PET scans (heart and knee prosthetic above, Thingiverse). The Micro3D software also allows advanced users to manipulate temperature, quality of the 3D print, and speed (Fig. 3.9).

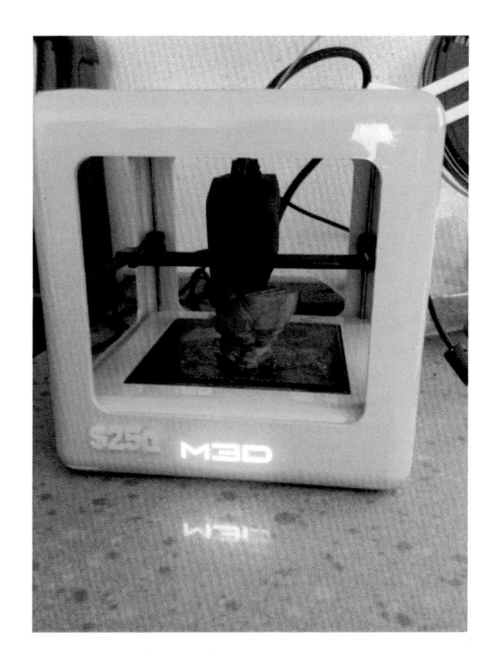

Figure 3.7 M3D SLA 3D printer photos in use with an HP PC Netbook.

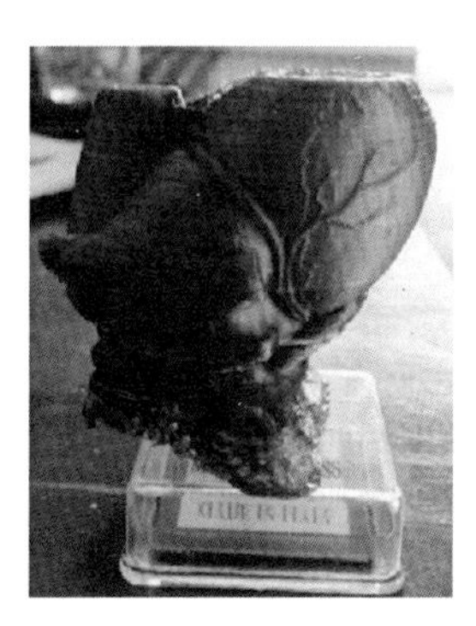

Figure 3.8 M3D SLA 3D printer photos in use with an HP PC Netbook.

iBox Nano (Cost in July 2015: USD $499)

iBox Nano 3D resin printer schematic and function per area of instrument (Fig. 3.10).

iBox Nano setup in my home

Slic3r File Converter Procedure for iBox Nano Printer

There is no website that has models formatted for the nano. All 3D prints I had to get and resolve in Slic3r and Meshmixer.

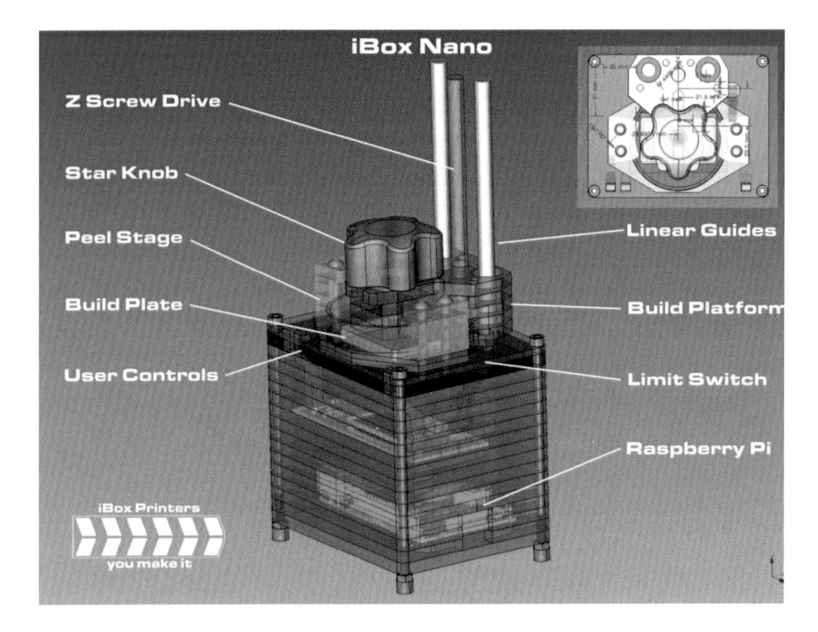

Figure 3.9 M3D SLA 3D printer photos in use with an HP PC Netbook.

Figure 3.10 M3D SLA 3D printer photos in use with an HP PC Netbook.

I was able to format all models to print on my iBox Nano.

Version of Slic3r used: 1.2.9a experimental

Procedure: Extract Slic3r to a directory of your choice and run the slic3r executable.

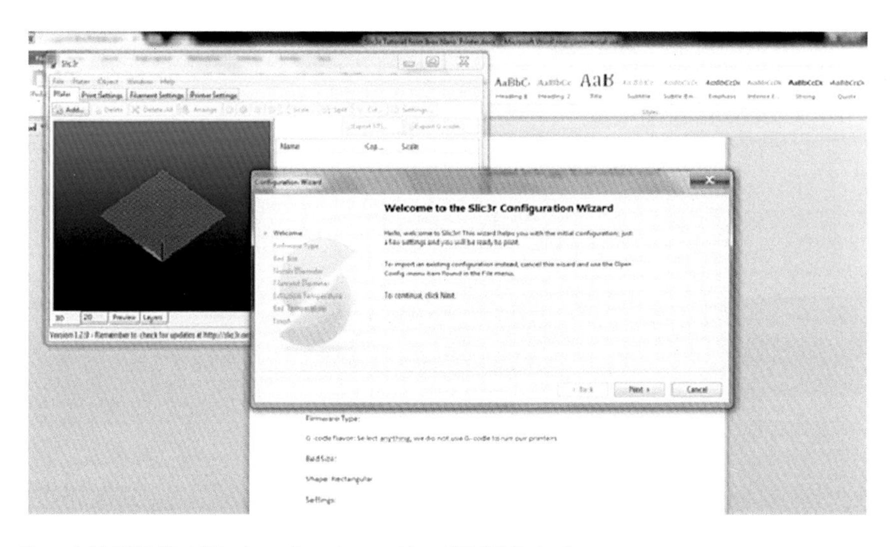

Figure 3.11 M3D SLA 3D printer photos in use with an HP PC Netbook.

First-time users may see the Configuration Wizard appear. If it doesn't show up automatically select Help- > Configuration Wizard... (Fig. 3.11).

Here are the settings with respect to each tab in the Configuration Wizard:

Firmware type: G-code flavor: Select anything, G-code is not used for the iBox Nano printer (Fig. 3.12).

Bed size:

Shape: Rectangular

Settings:

Size x: 40 y: 20

Origin x: 0 y: = 0 (Fig. 3.13)

Nozzle diameter: n/a

Not used, but the default is 0.5.

Filament diameter: n/a

Not used, but the default is 3

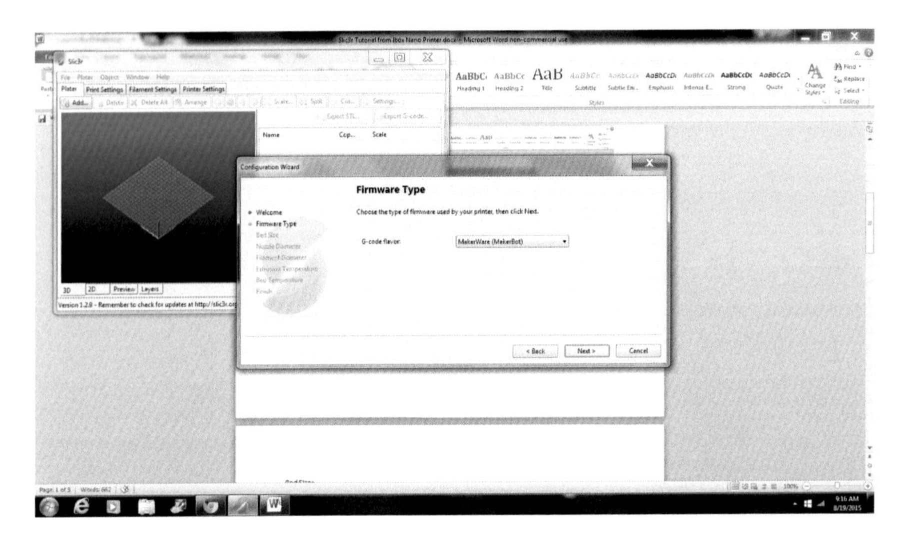

Figure 3.12 M3D SLA 3D printer photos in use with an HP PC Netbook.

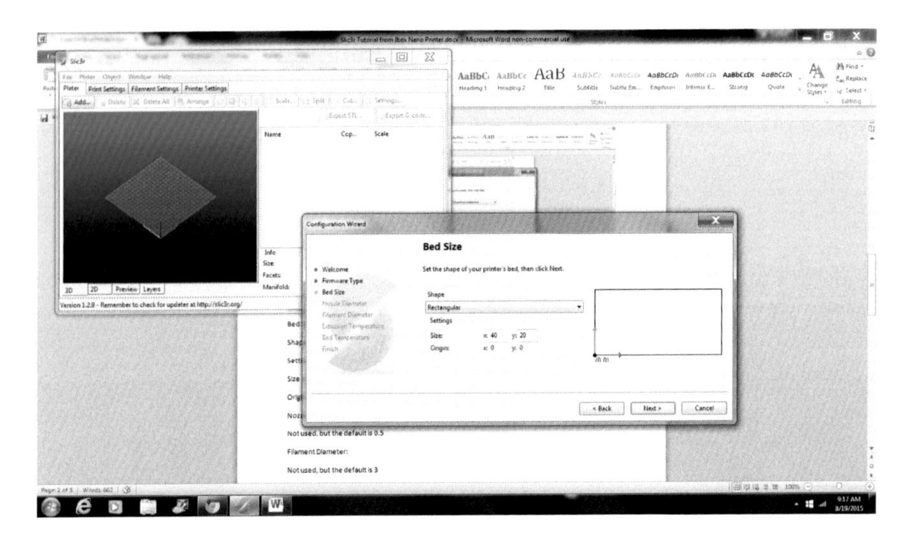

Figure 3.13 M3D SLA 3D printer photos in use with an HP PC Netbook.

Extrusion temperature: n/a

Not used, but the default is 200.

Bed temperature: n/a

Not used, but the default is 0 (Fig. 3.14)

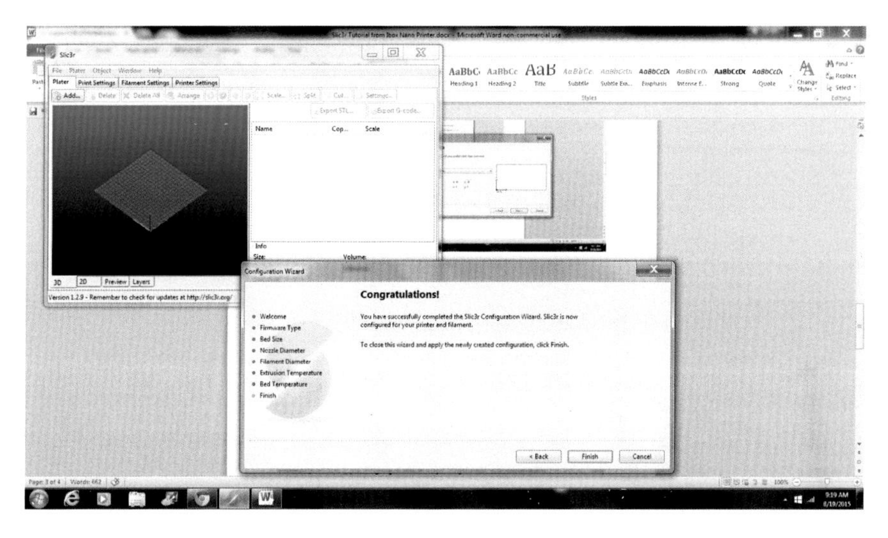

Figure 3.14 M3D SLA 3D printer photos in use with an HP PC Netbook.

After exiting the Configuration Wizard select the Print Settings on the main window.

Change the Layer height to 0.1 mm.

Note: iBox Nano 3D printers can support a minimum layer height of 2 um but the default of 100 um (0.1 mm) is recommended. You can change these settings to adjust print quality in the z direction at the cost of increased print time. If you are going to slice at a layer height other than the default of 100 um (0.1 mm) make sure you make the relevant changes in the Advanced Settings page of the iBox Nano. Exposure times are also going to change; thinner layers need shorter cure times and thicker layers need more.

Select the Platter Tab

Here I selected the add button to select a .stl model to slice (Figs. 3.15 and 3.16).

Once I selected the model and made sure it fits the boundaries of the build area after scaling and arranging it, I exported it to a new .stl file.

Now select File->Slice to SVG and select the .stl file you just created (Fig. 3.17).

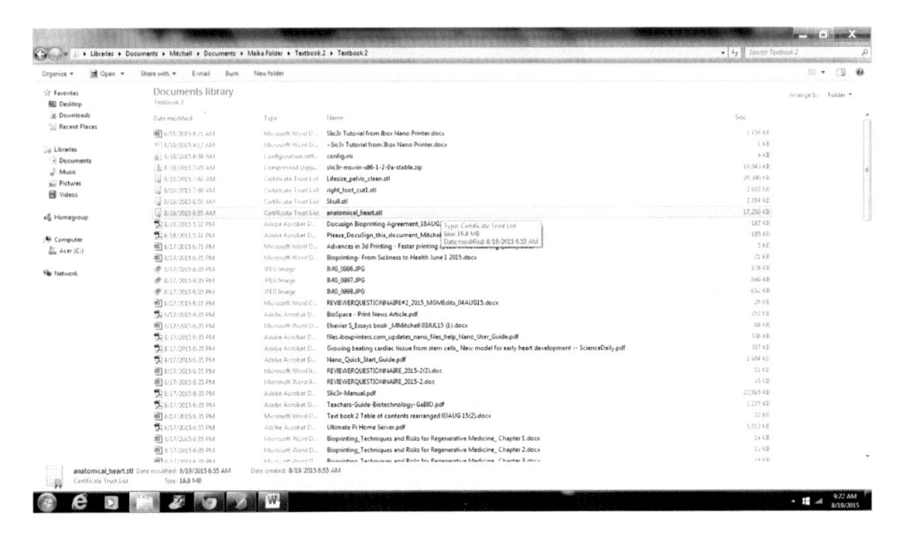

Figure 3.15 M3D SLA 3D printer photos in use with an HP PC Netbook.

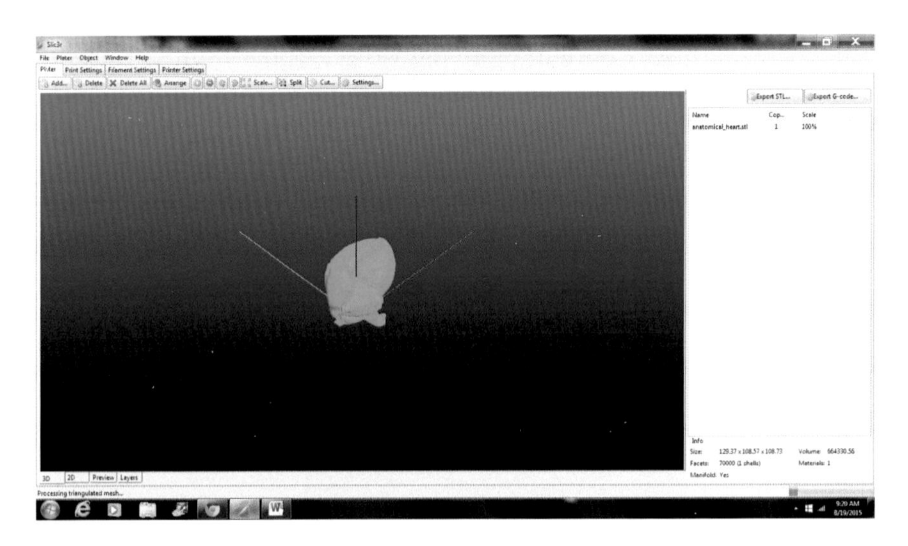

Figure 3.16 M3D SLA 3D printer photos in use with an HP PC Netbook.

Provide a location for slic3r to save the new SVG file (Figs. 3.18 and 3.19).

Once you slice your model you can use Gary Hodgson's SVG viewer to look at your .svg file once before uploading it to the iBox Nano resin printer (drag and drop the file onto the webpage).

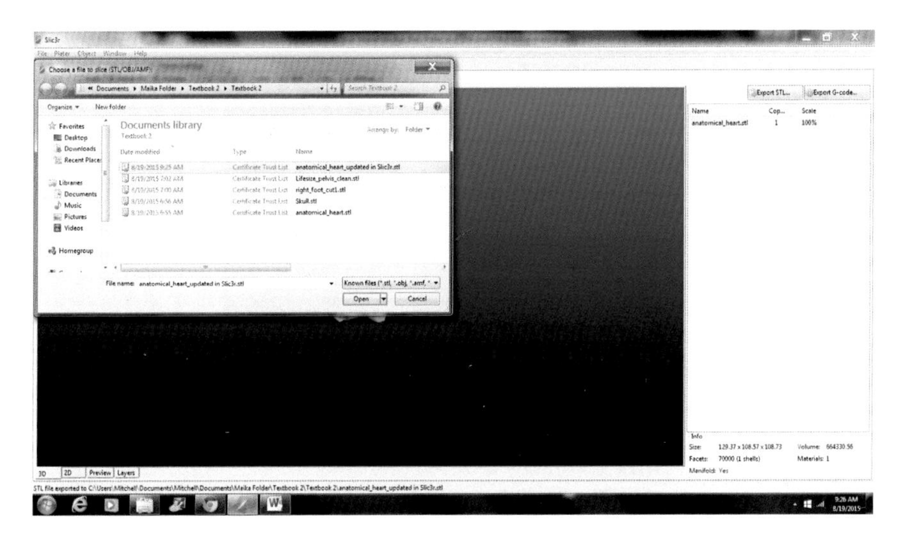

Figure 3.17 M3D SLA 3D printer photos in use with an HP PC Netbook.

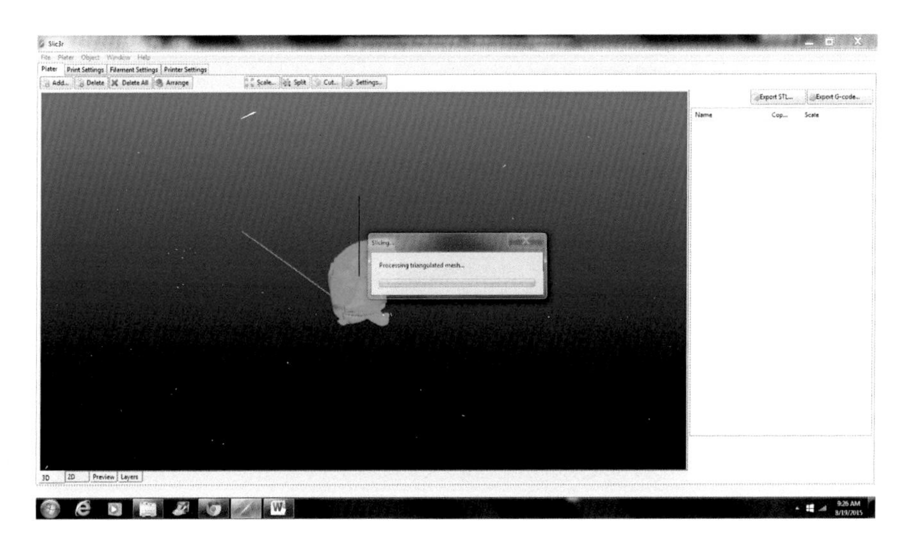

Figure 3.18 M3D SLA 3D printer photos in use with an HP PC Netbook.

I checked for the correct z height of the model because when slicing at 100 μm the number of layers it displays on the slider should be around 10 times the *z* height in mm (Figs. 3.20 and 3.21).

Upload the newly created .svg file to the iBox Nano printer using the + upload models button on the iBox Nano and follow the instructions onscreen (Figs. 3.22–3.27).

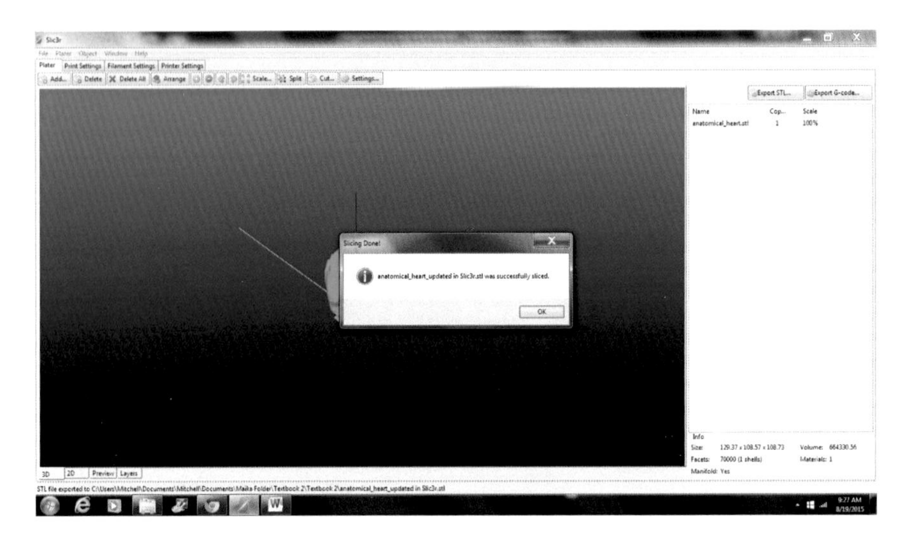

Figure 3.19 M3D SLA 3D printer photos in use with an HP PC Netbook.

Figure 3.20 M3D SLA 3D printer photos in use with an HP PC Netbook.

The first try didn't work. I realized I didn't scale the anatomical heart to the builder plate (Figs. 3.28 and 3.29). The heart printed successfully at this size.

I decided to update the file with increasing the size of the print by 190% while still fitting on the builder plate (Fig. 3.30):

Anatomical Heart pic A (Fig. 3.31)

Anatomical Heart pic A

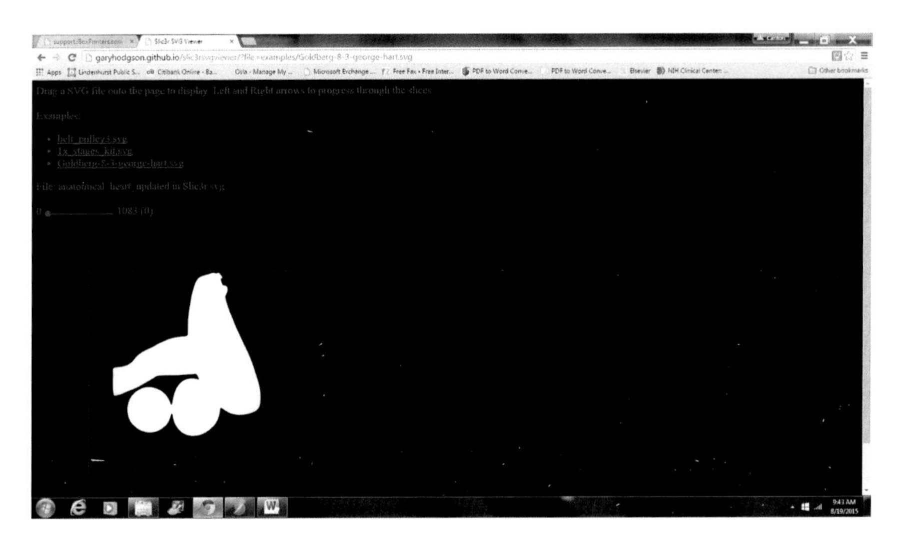

Figure 3.21 M3D SLA 3D printer photos in use with an HP PC Netbook.

Figure 3.22 M3D SLA 3D printer photos in use with an HP PC Netbook.

Preview of layers in A and B (Figs. 3.32–3.34).

Here is the final printed products using different color resins (Figs. 3.35–3.37).

This is the skull updated from *.stl to *.svg format as viewed in the iBox Nano preview (Figs. 3.38–3.43).

Figure 3.23 M3D SLA 3D printer photos in use with an HP PC Netbook.

Figure 3.24 M3D SLA 3D printer photos in use with an HP PC Netbook.

Due to the positioning of the rendering, raft/support material is added (which ultimately adds time to the original overall length of print time) in slic3r (Figs. 3.44–3.46).

If you are still unable to configure slic3r properly select File- > Load Config and load the config file attached in this post. (Windows users may need to run slic3r as an administrator to load the configuration.

Figure 3.25 M3D SLA 3D printer photos in use with an HP PC Netbook.

Figure 3.26 M3D SLA 3D printer photos in use with an HP PC Netbook.

To do this on Windows 7, right-click the slic3r executable and run as administrator. In other versions of Windows select Properties, click on the Compatibility tab, and check the box labeled Run as Administrator) (Figs. 3.47–3.50).

Learning to use the iBox Nano resin 3D printer was really challenging but exciting as it required a DIY mentality for the programming

Figure 3.27 M3D SLA 3D printer photos in use with an HP PC Netbook.

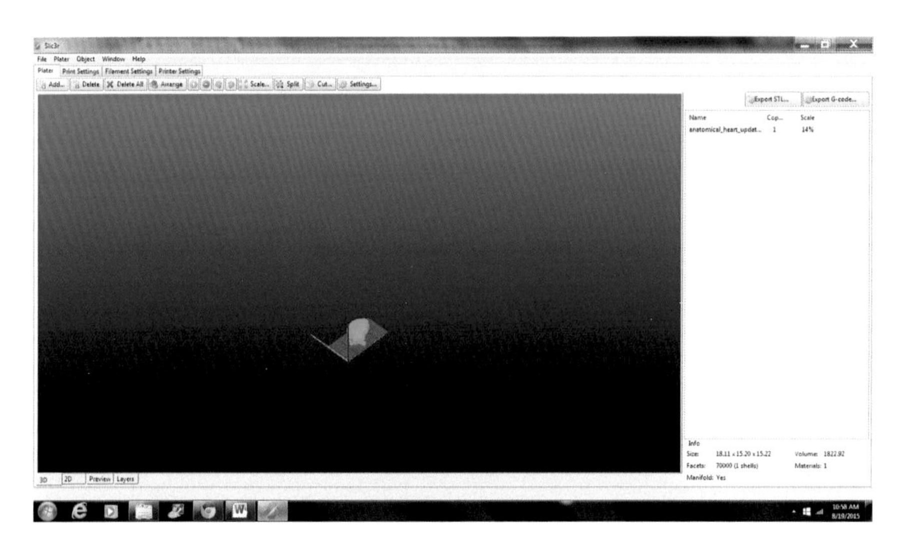

Figure 3.28 M3D SLA 3D printer photos in use with an HP PC Netbook.

and conversions in order to print. It definitely felt like a next level experience to the Micro3D, which is very much plug-and-play requiring little maintenance and programming. The resin printer can print the same exact objects in half the time using the same files (some conversion in filetypes is necessary between the Micro3D, which uses stereolithography and solid filament, compared to the iBox Nano, which utilizes liquid resin and UV to set each layer).

Figure 3.29 M3D SLA 3D printer photos in use with an HP PC Netbook.

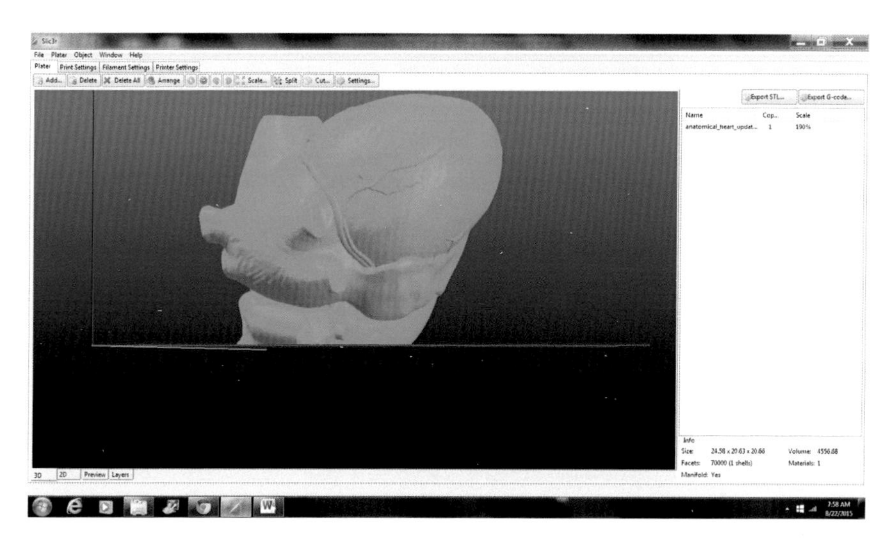

Figure 3.30 M3D SLA 3D printer photos in use with an HP PC Netbook.

Unfortunately, my utilization of the iBox Nano 3D printer came to an end as the toxicity of the resin was affecting my skin, even with gloves. There was also the issue of removal of waste and disposal.

FlashForge Dreamer (Cost in December 2015; USD $1300)

After my experience with the smaller 3D printers, I decided to buy a "maker" version with dual extruders, which allows you to use two

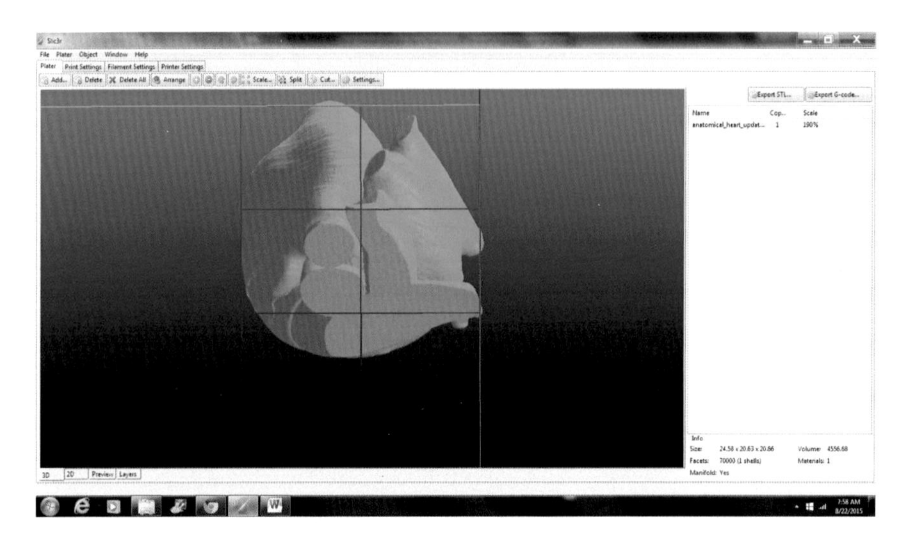

Figure 3.31 M3D SLA 3D printer photos in use with an HP PC Netbook.

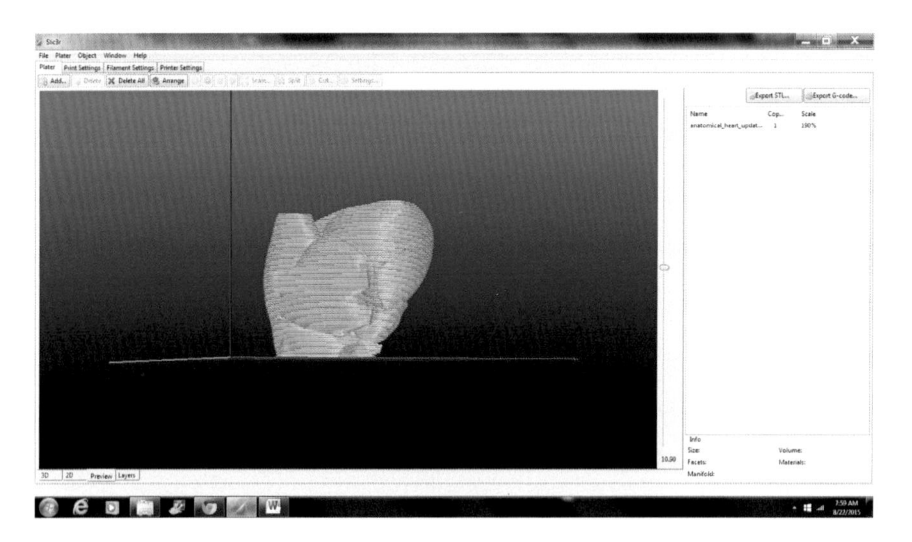

Figure 3.32 M3D SLA 3D printer photos in use with an HP PC Netbook.

different types of filaments and colors. The instrument came unassembled, so it required tools and my recent experience with 3D printers (and patience) (Fig. 3.51).

Here is a brief breakdown of the FlashForge Dreamer:

The Technique

Fused filament fabrication is the most common method of 3D printing. It is also the method that Dreamer uses. It works by melting

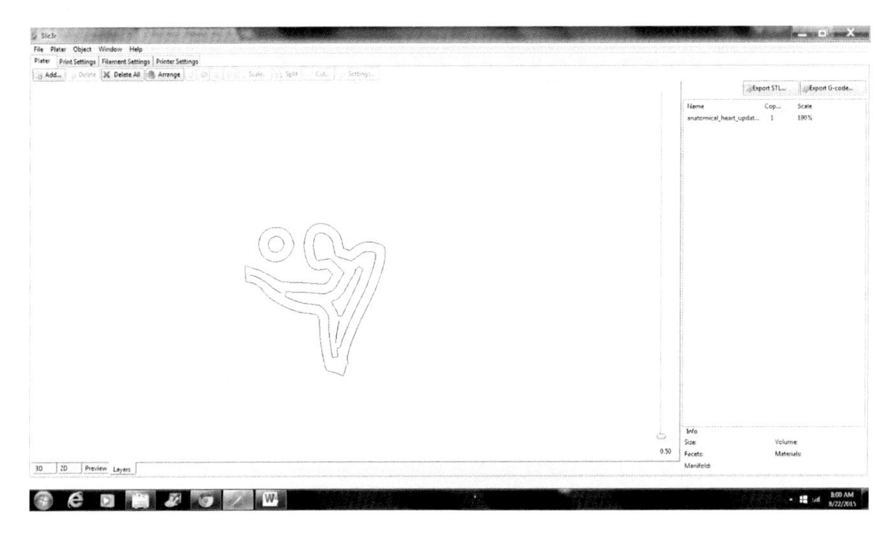

Figure 3.33 M3D SLA 3D printer photos in use with an HP PC Netbook.

Figure 3.34 M3D SLA 3D printer photos in use with an HP PC Netbook.

Figure 3.35 M3D SLA 3D printer photos in use with an HP PC Netbook.

Figure 3.36 M3D SLA 3D printer photos in use with an HP PC Netbook.

Figure 3.37 M3D SLA 3D printer photos in use with an HP PC Netbook.

plastic material called filament onto a print surface using high temperature. The filament solidifies after it cools down, which happens instantaneously after it is extruded from the print head. Three-dimensional objects are formed as the filament creates multiple layers. This particular printer can use all types of filament.

3D Printing Process

The 3D printing process involves three steps (Fig. 3.52):

1. 3D model design
2. Slicing and exporting the 3D model
3. Making the print

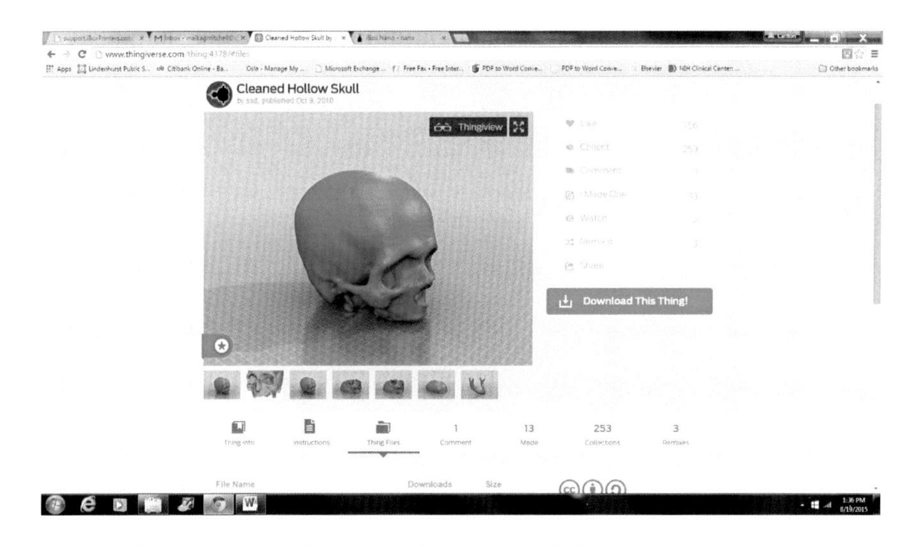

Figure 3.38 M3D SLA 3D printer photos in use with an HP PC Netbook.

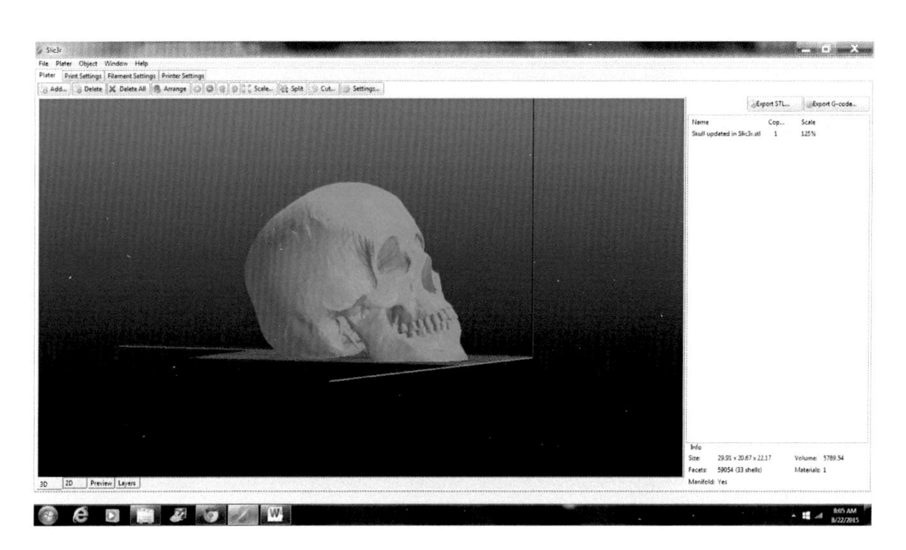

Figure 3.39 M3D SLA 3D printer photos in use with an HP PC Netbook.

MICRO 3D HEART 3D PRINT

(Fig. 3.53)

IBOX NANO HEART 3D PRINT

(Figs. 3.54 and 3.55)

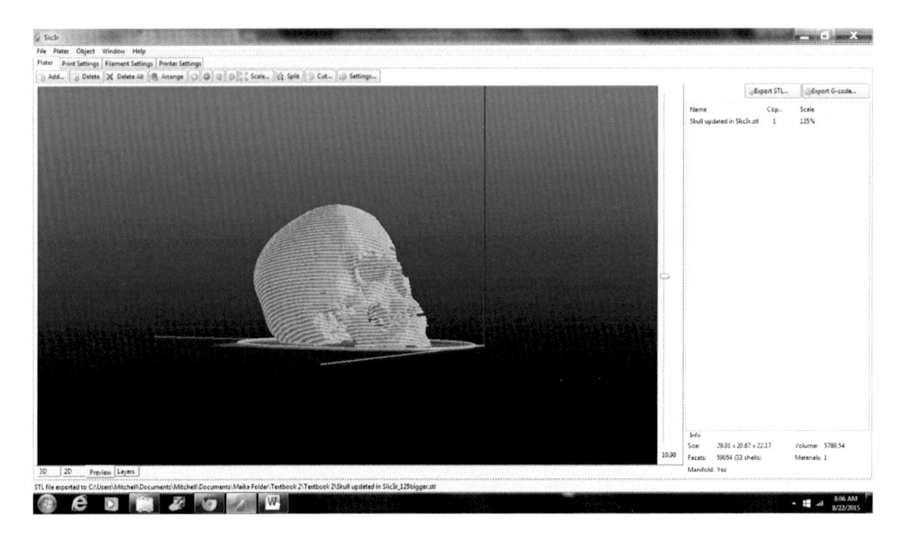

Figure 3.40 M3D SLA 3D printer photos in use with an HP PC Netbook.

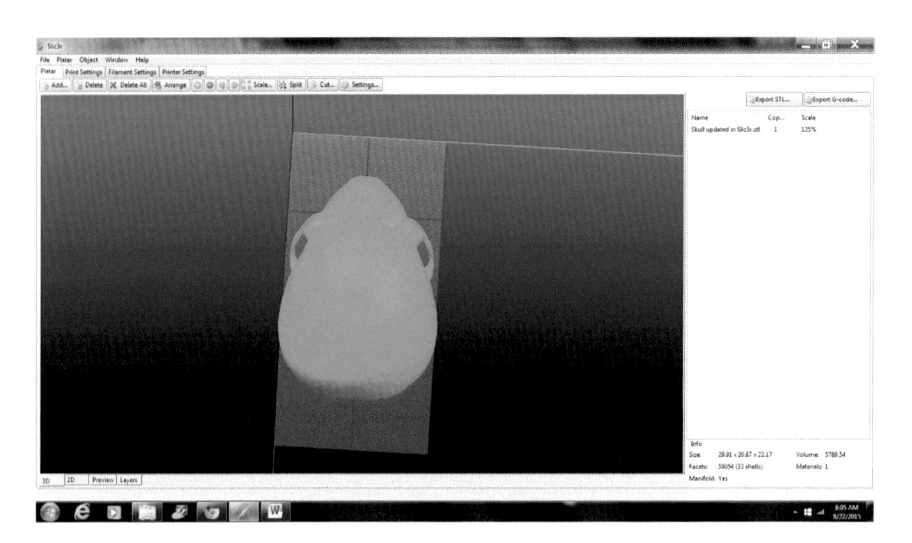

Figure 3.41 M3D SLA 3D printer photos in use with an HP PC Netbook.

FLASHFORGE DREAMER HEART 3D PRINT

Just as before, I used medical 3D files used for the Micro3D and iBox Nano.

To make a toy using this technique, a manufacturer loads a substance, usually plastic, into a mini-fridge-sized machine. He also loads

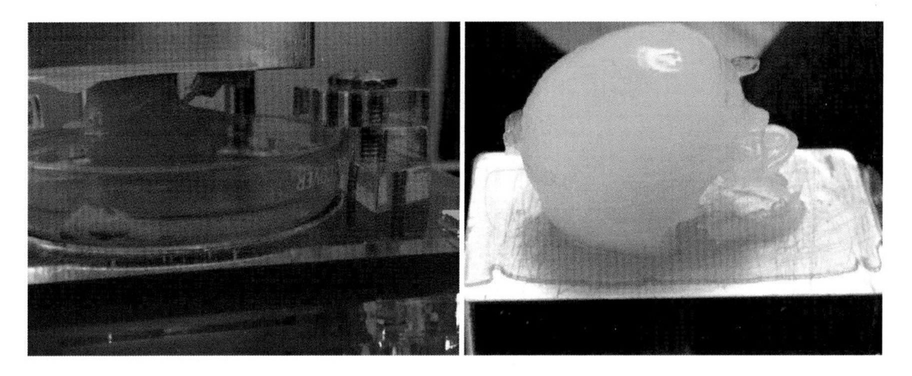

Figure 3.42 M3D SLA 3D printer photos in use with an HP PC Netbook.

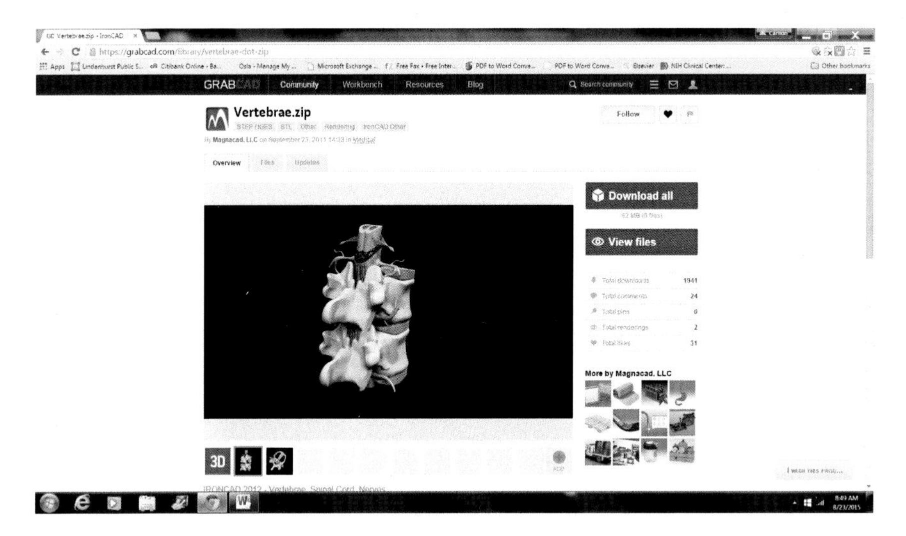

Figure 3.43 M3D SLA 3D printer photos in use with an HP PC Netbook.

a 3D design of the toy he wants to make. When he tells the machine to print, it heats up and, using the design as a set of instructions, extrudes a layer of melted plastic through a nozzle onto a platform. As the plastic cools, it begins to solidify, although by itself, it's nothing more than a single slice of the desired object. The platform then moves downward so a second layer can be deposited on the first. The printer repeats this process until it forms a solid object in the shape of the toy.

In industrial circles, this is known as *additive manufacturing* because the finished product is made by adding material to build up a 3D shape. It differs from traditional manufacturing, which often involves

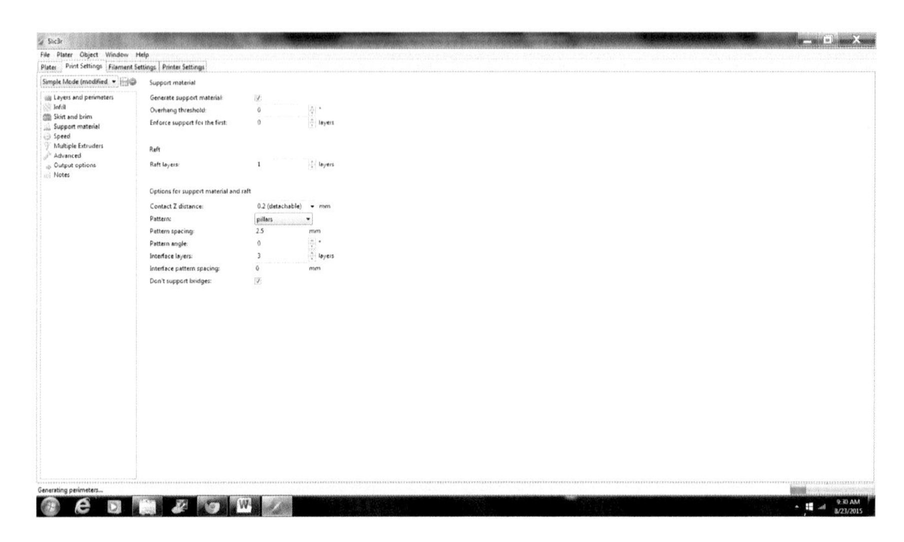

Figure 3.44 M3D SLA 3D printer photos in use with an HP PC Netbook.

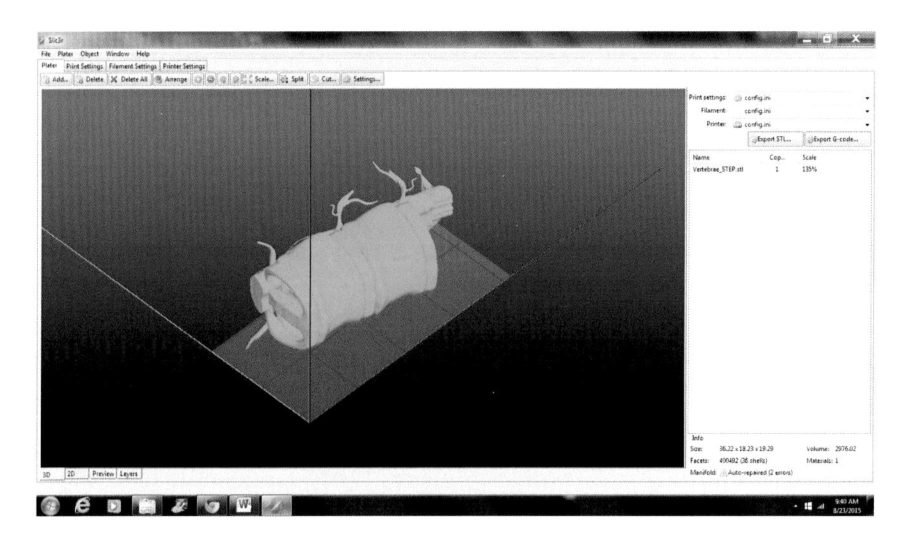

Figure 3.45 M3D SLA 3D printer photos in use with an HP PC Netbook.

subtracting a material, by way of machining, to achieve a certain shape. Additive manufacturers aren't limited to using plastic as their starting material. Some use powders, which are held together by glue or heated to fuse the powder together. Others prefer food materials, such as cheese or chocolate, to create edible sculptures. And still others—modern versions of Victor Frankenstein—are experimenting

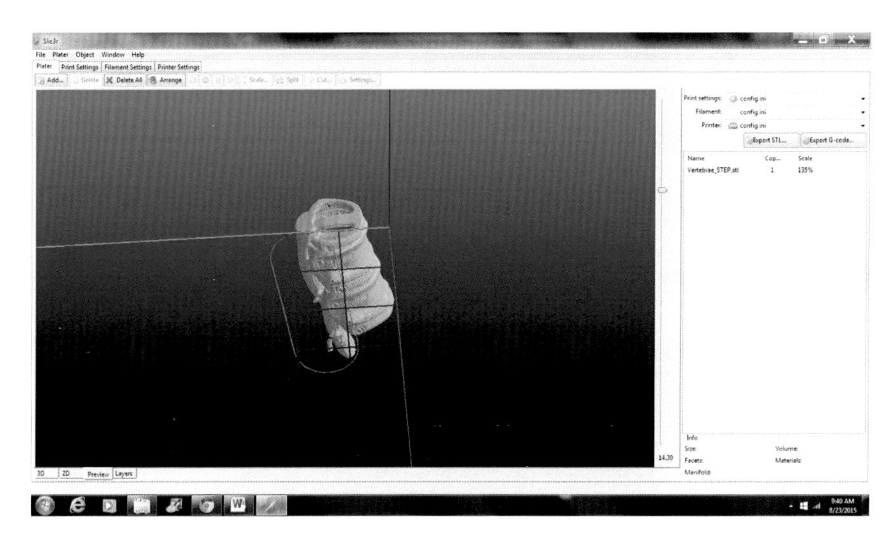

Figure 3.46 M3D SLA 3D printer photos in use with an HP PC Netbook.

Figure 3.47 M3D SLA 3D printer photos in use with an HP PC Netbook.

with biomaterials to print living tissue and, when layered properly in biotic environments, fully functioning organs.

That's right, the same technology that can produce Star Wars action figures can also produce human livers, kidneys, ears, blood

Figure 3.48 M3D SLA 3D printer photos in use with an HP PC Netbook.

Figure 3.49 M3D SLA 3D printer photos in use with an HP PC Netbook.

Figure 3.50 M3D SLA 3D printer photos in use with an HP PC Netbook.

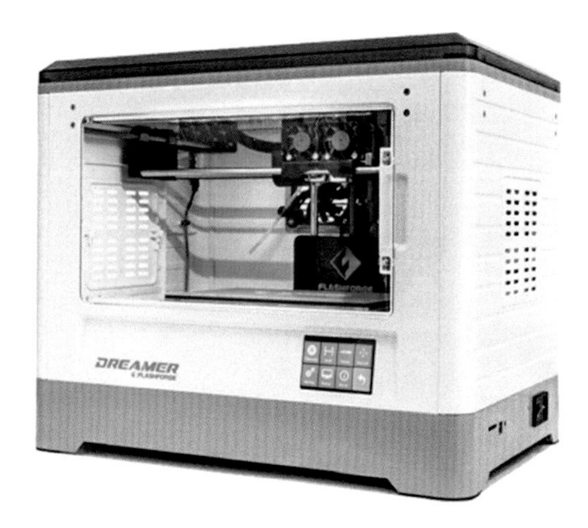

Figure 3.51 M3D SLA 3D printer photos in use with an HP PC Netbook.

Figure 3.52 M3D SLA 3D printer photos in use with an HP PC Netbook.

Figure 3.53 M3D SLA 3D printer photos in use with an HP PC Netbook.

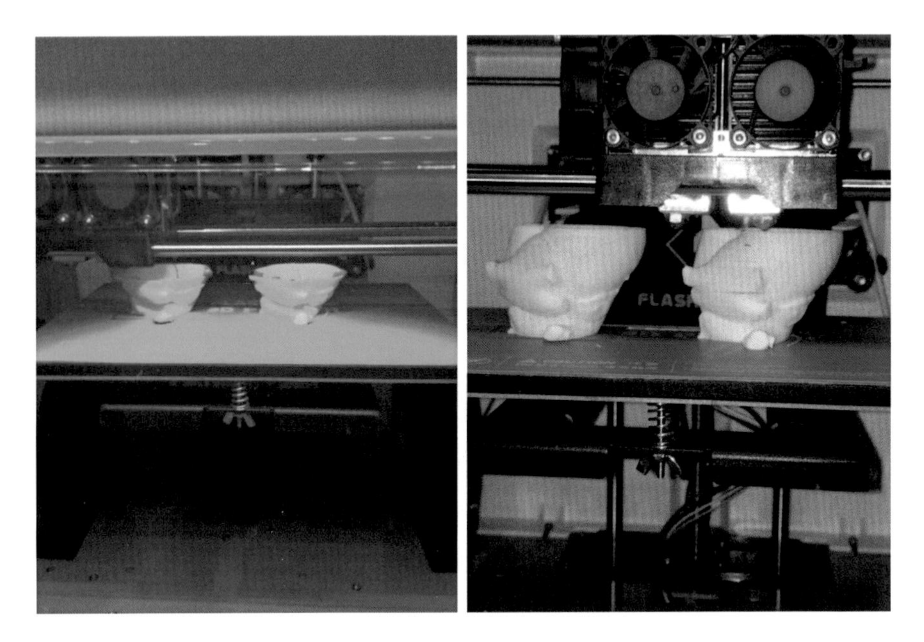

Figure 3.54 M3D SLA 3D printer photos in use with an HP PC Netbook.

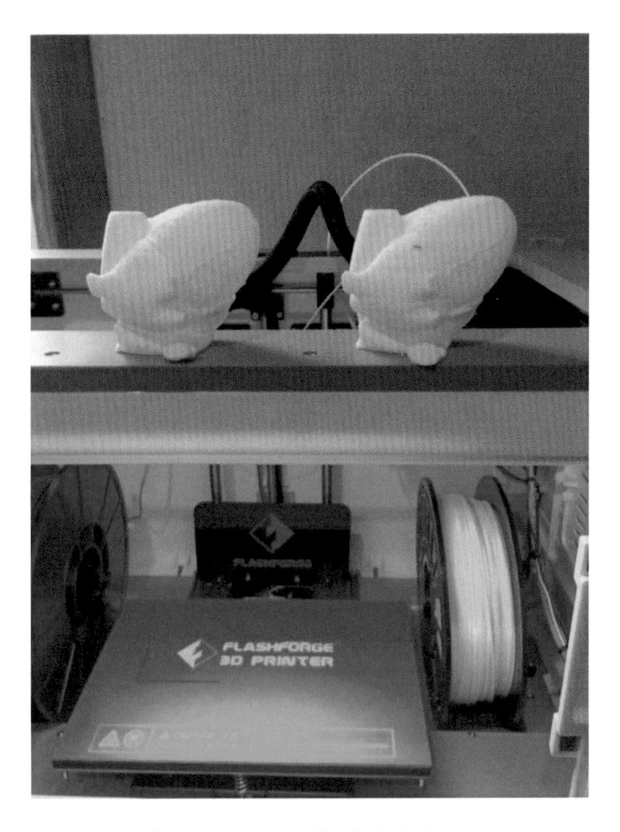

Figure 3.55 M3D SLA 3D printer photos in use with an HP PC Netbook.

vessels, skin, and bones. But printing a 3D version of R2-D2 isn't exactly the same as printing a heart that expands and contracts like real cardiac muscle. Cut through an action figure, and you'll find plastic through and through. Cut through a human heart, and you'll find a complex matrix of cells and tissues, all of which must be arranged properly for the organ to function. For this reason, bioprinting is developing more slowly than other additive manufacturing techniques, but it is advancing. Researchers have already built modified 3D printers and are now perfecting the processes that will allow them to print tissues and organs for pharmaceutical testing and, ultimately, for transplantation.

THE 3D HISTORY OF BIOPRINTING

The promise of printing human organs began in 1983 when Charles Hull invented stereolithography. This special type of printing used a laser to solidify a polymer material extruded from a nozzle. The instructions for the design came from an engineer, who would define the 3D shape of an object in CAD software and then send the file to the printer. Hull and his colleagues developed the file format, known as .stl, that carried information about the object's surface geometry, represented as a set of triangular faces.

At first, the materials used in stereolithography weren't sturdy enough to create long-lasting objects. As a result, engineers in the early days used the process strictly as a way to model an end product—a car part, for example—that would eventually be manufactured using traditional techniques. An entire industry, known as rapid prototyping, grew up around the technology, and in 1986, Hull founded 3D Systems to manufacture 3D printers and the materials to go in them.

By the early 1990s, 3D Systems had begun to introduce the next generation of materials—*nanocomposites*, blended plastics and powdered metals. These materials were more durable, which meant they could produce strong, sturdy objects that could function as finished products, not mere stepping stones to finished products.

It didn't take long for medical researchers to notice. What's an organ but an object possessing a width, height, and depth? Couldn't such a structure be mapped in three dimensions? And couldn't a 3D printer receive such a map and then render the organ the same way it might render a hood ornament or piece of jewelry? Such a feat could be easily accomplished if the printer cartridges sprayed out biomaterials instead of plastics.

Scientists went on the hunt for such materials and by the late 1990s they had devised viable techniques and processes to make organ-building a reality. In 1999, scientists at the Wake Forest Institute for Regenerative Medicine used a 3D printer to build a synthetic scaffold of a human bladder. They then coated the scaffold with cells taken from their patients and successfully grew working organs. This set the stage for true bioprinting. In 2002, scientists printed a miniature functional kidney capable of filtering blood and producing urine in an animal model. And in 2010, Organovo—a bioprinting company headquartered in San Diego—printed the first blood vessel.

JUST LIKE AN INKJET PRINTER, SORT OF

The idea of 3D printing evolved directly from a technology everyone knows: the inkjet printer. Watch your HP or Epson machine churn out a printed page, and you'll notice that the print head, driven by a motor, moves in horizontal strips across a sheet of paper. As it moves, ink stored in a cartridge sprays through tiny nozzles and falls on the page in a series of fine drops. The drops build up to create an image, with higher-resolution settings depositing more ink than lower-resolution settings. To achieve full top-to-bottom coverage, the paper sheet, located beneath the print head, rolls up vertically.

The limitation of inkjet printers is that they only print in two dimensions—along the x- and y-axes. A 3D printer overcomes this by adding a mechanism to print along an additional axis, usually labeled the z-axis in mathematical applications. This mechanism is an elevator that moves a platform up and down. With such an arrangement, the ink head can lay down material from side to side, but it can also deposit layers vertically as the elevator draws the platform down and away from the print head. Fill the cartridge with plastic, and the printer will output a 3D plastic widget. Fill it with cells, and it will output a mass of cells.

Conceptually, bioprinting is really that simple. In reality, it's a bit more challenging because an organ contains more than one type of material. And because the material is living tissue, it needs to receive nutrients and oxygen. To accommodate this, bioprinting companies have modified their 3D printers to better serve the medical community.

Bioprinter Basic Parts

Print Head Mount—On a bioprinter, the print heads are attached to a metal plate running along a horizontal track. The x-axis motor propels the metal plate (and the print heads) from side to side, allowing material to be deposited in either horizontal direction.

Elevator—A metal track running vertically at the back of the machine, the elevator, driven by the z-axis motor, moves the print heads up and down. This makes it possible to stack successive layers of material, one on top of the next.

Platform—A shelf at the bottom of the machine provides a platform for the organ to rest on during the production process. The platform may support a scaffold, a petri dish, or a well plate, which could contain up to 24 small depressions to hold organ-tissue samples for pharmaceutical testing. A third motor moves the platform front to back along the y-axis.

Reservoirs—The reservoirs attach to the print heads and hold the biomaterial to be deposited during the printing process. These are equivalent to the cartridges in your inkjet printer.

Print Heads/Syringes—A pump forces material from the reservoirs down through a small nozzle or syringe, which is positioned just above the platform. As the material is extruded, it forms a layer on the platform.

Triangulation Sensor—A small sensor tracks the tip of each print head as it moves along the x-, y-, and z-axes. Software communicates with the machine so the precise location of the print heads is known throughout the process.

Microgel—Unlike the ink you load into your printer at home, bioink is alive, so it needs food, water, and oxygen to survive. This nurturing environment is provided by a microgel—think gelatin enriched with vitamins, proteins, and other life-sustaining compounds. Researchers either mix cells with the gel before printing or extrude the cells from one print head, microgel from the other. Either way, the gel helps the cells stay suspended and prevents them from settling and clumping.

Bioink—Organs are made of tissues, and tissues are made of cells. To print an organ, a scientist must be able to deposit cells specific to the organ she hopes to build. For example, to create a liver, she would start with hepatocytes—the essential cells of a liver—as well as other supporting cells. These cells form a special material known as *bioink*, which is placed in the reservoir of the printer and then extruded through the print head. As the cells accumulate on the platform and become embedded in the microgel, they assume a 3D shape that resembles a human organ.

Alternatively, the scientist could start with a bioink consisting of stem cells, which, after the printing process, have the potential to differentiate into the desired target cells. Either way, bioink is simply a medium, and a bioprinter is an output device. Up next, we'll review the steps required to print an organ designed specifically for a single patient.

The heart may be one of the easier organs to make with a bioprinter, stated Stuart K. Williams, the head of the Cardiovascular Innovation Institute, in a 2013 interview with Wired.

STEPS TO PERSONALIZED MEDICINE; MADE-TO-ORDER HUMAN ORGANS

When researchers built 3D printers capable of depositing bioink and forming living masses of cells, they celebrated a major achievement. Then they immediately began to tackle the next big problem: How can bioprinting produce an organ for a specific person? To accomplish this, a medical team needs to collect data about the organ in question—its size, shape, and placement in the patient's body. Then team members need to concoct a bioink using cells taken from the patient. This ensures that the printed organ will be compatible genetically and won't be rejected once it's transplanted in the patient's body.

For simple organs, such as bladders, researchers don't print the living tissue directly. Instead, they print a 3D scaffold made of biodegradable polymers or collagen. To determine the exact shape of the scaffold, they first build a 3D model using CAD software. They usually define the exact *x*-, *y*-, and *z*-coordinates of the model by taking scans of the patient using CT or MRI technology.

Next, researchers get the cells they need by taking a biopsy of the patient's bladder. They then place the cell samples in a culture, where they multiply into a population sufficiently large enough to cover the scaffold, which provides a temporary substrate for the cells to cling to as they organize and strengthen. Seeding the scaffold requires time-consuming and painstaking handwork with a pipette. It generally takes about 8 weeks before such artificial bladders are ready for implantation. When doctors finally place the organ in the patient, the scaffold has either disappeared or disappears soon after the surgery.

The procedure above works because bladder tissue only contains two types of cells. Organs like kidneys and livers have a far more complex structure with a greater diversity of cell types. While it would be easy enough to print a scaffold, it would be almost impossible to recreate the 3D structure of the tissue manually. A bioprinter, however, is ideally suited to complete such a time-consuming, detail-oriented task.

CONVENTIONAL STEPS PROPOSED FOR 3D BIOPRINTING HUMAN ORGANS

Here are the steps to print a complex organ:

- First, doctors make CT or MRI scans of the desired organ.
- Next, they load the images into a computer and build a corresponding 3D blueprint of the structure using CAD software.
- Combining this 3D data with histological information collected from years of microscopic analysis of tissues, scientists build a slice-by-slice model of the patient's organ. Each slice accurately reflects how the unique cells and the surrounding cellular matrix fit together in 3D space.
- After that, it's a matter of hitting File > Print, which sends the modeling data to the bioprinter.
- The printer outputs the organ one layer at a time, using bioink and gel to create the complex multicellular tissue and hold it in place.
- Finally, scientists remove the organ from the printer and place it in an incubator, where the cells in the bioink enjoy some warm, quiet downtime to start living and working together. For example, liver cells need to form what biologists call "tight junctions," which

describes how the cell membrane of one cell fuses to the cell membrane of the adjacent cell. The time in the incubator really pays off—a few hours in the warmth turns the bioink into living tissue capable of carrying out liver functions and surviving in a lab for up to 40 days.

The final step of this process—making printed organ cells behave like native cells—has been challenging. Some scientists recommend that bioprinting be done with a patient's stem cells. After being deposited in their required 3D space, they would then differentiate into mature cells, with all of the instructions about how to "behave." Then, of course, there's the issue of getting blood to all of the cells in a printed organ. Currently, bioprinting doesn't offer sufficient resolutions to create tiny, single-cell-thick capillaries. But scientists have printed larger blood vessels, and as the technology improves, the next step will be fully functional replacement organs, complete with the vascularization necessary to remain alive and healthy.

In the United States, that role would fall to the Food and Drug Administration (FDA), but this is new territory for the FDA as well. According to an August 2013, blog post, two labs in the agency's Office of Science and Engineering Laboratories (OSEL) are on the case. The Laboratory for Solid Mechanics is busy evaluating "how different printing techniques and processes affect the strength and durability of the materials used in medical devices." The Functional Performance and Device Use Laboratory has "developed and adapted computer-modeling methods to help us determine the effect of design changes on the safety and performance of devices when used in different patient populations" (source: Pollack and Coburn).

USES FOR 3D ORGANS

At the time of writing, surgeons have not implanted an organ printed from scratch into a human. That doesn't mean there haven't been successes. Replacing parts of the skeleton is one area being revolutionized by 3D printing. Some dentists now take an intraoral scan of a patient's teeth and send the scan to a lab that fashions a porcelain bridge using a 3D printer. Prosthetic manufacturers have also changed their approach to designing artificial limbs. Today, many are able to print fairings—prosthetic limb covers—that mold perfectly to a person's anatomy, giving the wearer a more comfortable fit. These are just preludes to what the future may hold: printing entire bones for placement in the body. Doctors in The Netherlands have already created a lower mandible on a 3D printer and implanted the jaw—made from bioceramic-coated titanium—in a patient suffering from a chronic bone infection.

Scientists have also successfully printed cartilaginous structures, such as ears and tracheas. To make the former, bioengineers take a 3D scan of a patient's ear, design a mold using CAD software, and then print it out. Then they inject the mold with cartilage cells and collagen. After spending some time in an incubator, the ear comes out, ready for attachment to the patient. A trachea can be made in a similar fashion. In 2012, doctors at the University of Michigan printed a sleeve, made from a 3D model generated from a CT scan, to wrap and support a baby's trachea, which had been rendered weak and floppy by a rare defect.

The holy grail, of course, is a bioprinted organ, and skin—the body's largest organ—may be the first item on the list. Researchers at the Wake Forest Institute for Regenerative Medicine already have developed a complete system to print skin grafts. The system includes a scanner to map a patient's wound and a purpose-built inkjet printer that lays down the cells, proteins, and enzymes necessary to form human skin. The goal is to build portable printers for use in field hospitals, where doctors can output skin directly onto patients.

Until these marvels come online, 3D organs will play an important role in education and drug development. They might even factor into the development of food and clothing products (lab-grown meat and leather). Some medical schools have invested in 3D printing technology to create surgical models of organs from CT or MRI images. This allows students to practice on hearts, livers, and other structures that look and feel just like the real thing. Having access to such lifelike tissues also benefits pharmaceutical companies, which can test candidate drugs to see their effects. Organovo houses several printers capable of printing out 3D models of liver, kidney, and cancer tissues. These aren't full organs meant to live indefinitely. Instead, they're "organs on a chip"—small, biologically active tissue samples designed to respond as native tissues would.

I currently have a provisional patent approved for a new type of biofilament to be used on the flashforge printer for bioscaffolding for bone.

FURTHER READING

[1] Atala, A. Printing a human kidney. TED Talks. March 2011. (Nov. 17, 2013) <http://www.ted.com/talks/anthony_atala_printing_a_human_kidney.html>.

[2] Banham, R. Printing a medical revolution. T. Rowe Price. May 2012. (Nov. 17, 2013) <http://individual.troweprice.com/public/Retail/Planning-&-Research/Connections/3D-Printing/Printing-a-Medical-Revolution>.

[3] Clark, L. Bioengineer: the heart is one of the easiest organs to bioprint, we'll do it in a decade. Cardiovascular Innovation Institute. Nov. 21, 2013. (Nov. 25, 2013) <http://cv2i.org/bioengineer-heart-one-easiest-organs-bioprint-well-decade/>.

[4] Dutton, G. 3D printing may revolutionize drug R&D. Genetic Engineering & Biotechnology News. Nov. 15, 2013. (Nov. 17, 2013) <http://www.genengnews.com/gen-articles/3D-printing-may-revolutionize-drug-r-d/5062/>.

[5] Fountain, H. At the printer, living tissue. The New York Times. Aug. 18, 2013. (Nov. 17, 2013) <http://www.nytimes.com/2013/08/20/science/next-out-of-the-printer-living-tissue.html?>.

[6] History.com Staff. Organ transplants: a brief history. History.com. Feb. 21, 2012. (Nov. 17, 2013) <http://www.history.com/news/organ-transplants-a-brief-history>.

[7] Hsu, J. 3D printed organs may mean end to waiting lists, deadly shortages. Huffington Post. Sept. 25, 2013. (Nov. 17, 2013) <http://www.huffingtonpost.com/2013/09/25/3D-printed-organs_n_3983971.html>.

[8] Hsu, J. 3D printing: what a 3D printer is and how it works. LiveScience. May 21, 2013. (Nov. 17, 2013) <http://www.livescience.com/34551-3D-printing.html>.

[9] Hsu, J. Tiny 3D-printed organs aim for 'body on a chip'. LiveScience. Sept. 16, 2013. (Nov. 17, 2013) <http://www.livescience.com/39660-3D-printed-body-on-a-chip.html>.

[10] Image Specialists. How inkjet printers work. <http://www.image-specialists.com/ink_int_injet_printer.aspx>.

[11] Leckart, S. The body shop, popular science. Popular Science. Aug. 2013.

[12] Organovo. The bioprinting process. (Nov. 17, 2013) <http://www.organovo.com/science-technology/bioprinting-process>.

[13] Pollack, S.K., Coburn, J. FDA goes 3D. FDAVoice. Aug. 15, 2013. (Dec. 13, 2013) <http://blogs.fda.gov/fdavoice/index.php/2013/08/fda-goes-3-d/>.

[14] Royte, E. What lies ahead for 3D printing? Smithsonian Magazine. May 2013. (Nov. 17, 2013) <http://www.smithsonianmag.com/science-nature/What-Lies-Ahead-for-3D-Printing-204136931.html>.

[15] Ungar, L. Researchers closing in on printing 3D hearts. USA Today. May 29, 2013. (Nov. 17, 2013) <http://www.usatoday.com/story/tech/2013/05/29/health-3D-printing-organ-transplant/2370079/>.

[16] Wake Forest Institute for Regenerative Medicine. Using ink-jet technology to print organs and tissue. Aug. 2, 2013. (Nov. 17, 2013) <http://www.wakehealth.edu/Research/WFIRM/Our-Story/Inside-the-Lab/Bioprinting.html>.

CHAPTER 4

Bioprinters in Use Today

Bioprinters: Organovo released the NovoGen MMX, the world's first production 3D bioprinter. The printer has two robotic printheads: One places human cells and the other places a hydrogel, scaffold, or other type of support. There have also been other pioneers in this field, which began over 40 years ago as shown in Tables 4.1 and 4.2.

INKJET-BASED BIOPRINTING

In inkjet-based bioprinting (Table 4.2; Fig. 4.1), a bioink made of cells and biomaterials are printed in the form of droplets through cartridges onto a substrate. There are two types of inkjet printing: continuous inkjet printing (CIJ) and drop-on-demand inkjet printing (DOD). In CIJ mode, a jet is obtained by forcing the liquid through an orifice under an external pressure, which breaks it up into a stream of droplets. In DOD mode, a pressure pulse is applied to the fluid, which generates drops only when needed. For inkjet printing of cells, thermal and piezo-electric inkjet printing are the two most commonly used methods. For thermal inkjet printing, small volumes of the printing fluid are vaporized by a micro-heater to create the pulse that expels droplets from the printhead. In piezoelectric inkjet printing, a direct mechanical pulse is applied to the fluid in the nozzle by a piezoelectric actuator, which causes a shock wave that forces the bioink through the nozzle [10,15,40,47]. However, there have been only a few examples of cell deposition by piezoelectric inkjet printing due to the electrically conducting ink formulations and contamination concerns on ink recycling [24].

Inkjet-based bioprinting (Fig. 4.1) allows different cell types to be deposited in precise orientations relative to the print surface and to each other at micrometer resolution by controlling the ejection nozzles and timing of spray [48]. Wilson and Boland first adapted inkjet printers for the manufacture of cell and protein arrays, which have the advantage of being fully automated and computer controlled [49]. In

Bioprinting. DOI: http://dx.doi.org/10.1016/B978-0-12-805369-0.00004-3

Table 4.1 A Brief History of the Bioprinter
1976—The inkjet printer is invented.
1984—The 3D printer is invented and patented by physicist Charles Hull (US 4575330 A). Technically termed stereolithography, Hull's invention allows a physical object to be printed from a 3D computer model.
1990s—Inkjet printers achieve a resolution of 80 μm, opening the door for bioprinter conversion.
1999—An artificially created bladder is successfully transplanted into a beagle, regular urinary function is observed within a month [1].
2002—A small, working kidney is 3D printed in the lab [1].
2008—A 3D printed prosthetic leg is implanted in a human. All parts were printed together, with no assembly required.
2012—Miller et al. successfully 3D print permeable blood vessels using an open-source 3D printer and biodegradable glass channels [2].
2014—Kolesky et al. successfully create an interwoven vascular network using novel Pluronic F127 fugitive ink [3].

Table 4.2 Popular Commercial Bioprinters in Use [2–16,28,31,36–39,50,51,54,74–77,79,80,82,83,90–189]

	Inkjet	Microextrusion	Laser assisted
Gelation methods	Chemical, photo cross-linking	Chemical, photo-cross-linking, sheer thinning, temperature	Chemical, photo cross-linking
Print speed	Fast (1–10,000 droplets/s)	Slow (10–50 μm/s)	Medium-fast (200–1600 mm/s)
Cell viability	>85%	40–80%	>95%
Cell plating densities	Low, <10^6 cells/mL	High, cell spheroids	Medium, 10^8 cells/mL
Printer cost + consumables	$	$$$	>$$$$

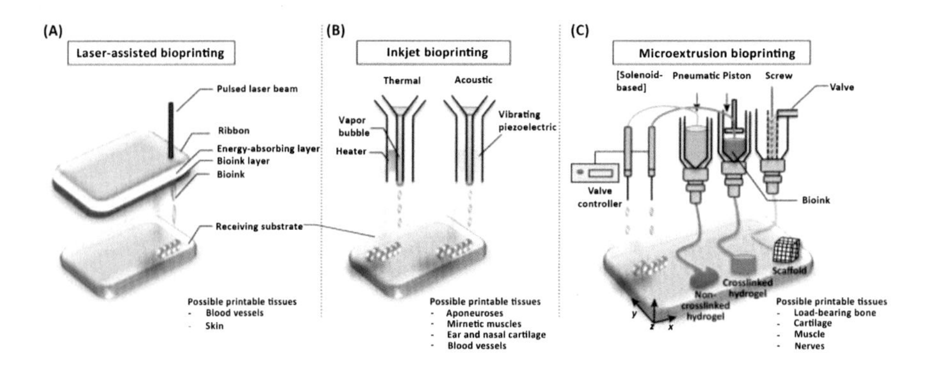

Figure 4.1 Current types of bioprinters in the market (2016).

their next study, cell aggregates were printed onto thermosensitive gels layer-by-layer in order to demonstrate fusion between the closely placed cell aggregates [50]. The same group deposited Chinese hamster ovary (CHO) cells and rat embryonic motoneurons as bioink.

Although inkjet bioprinting is one of the most commonly used methods for printing living cells and biomaterials, cell aggregation, sedimentation, and cell damage due to high shear stresses are some of the disadvantages of this method. Cell aggregation and sedimentation may be prevented by frequent stirring of the cell mixture, which can result in reduced cell viability if the cells are sensitive to the shear forces [57]. Another problem limiting inkjet bioprinting is the clogging of the nozzle orifice. Low viscosity surfactants can be added to the ink but these can create additional challenges such as cell damage [58].

LASER-ASSISTED BIOPRINTING

While laser-based approaches (Table 4.2; Fig. 4.1) allow researchers to pattern living cells on a substrate [29] and to layer multiple cell types [30], laser-based techniques have also been explored for positioning of cells in microarrays [31]. The resolution of laser-assisted bioprinting is affected by different factors such as the laser fluence, the wettability of the substrate, and the thickness and viscosity of the biological layer [32]. Guillotion and his group studied the effect of the viscosity of the bioink, laser energy, and laser printing speed on the resolution of cell printing [33].

Organovo

At Organovo, functional human tissues using proprietary 3D bioprinting technology are created. Their goal is to build living human tissues that can function like native tissues. They have been able to accomplish reproducible 3D tissues that represent human biology, and are enabling groundbreaking therapies by:

- Partnering with biopharmaceutical companies and academic medical centers to design, build, and validate more predictive in vitro tissues for disease modeling and toxicology.
- Creating opportunities to test drugs on functional human tissues before ever administering the drug to a living person, bridging the gulf between preclinical testing and clinical trials.
- Providing functional, 3D tissues that can be implanted or delivered into the human body to repair or replace damaged or diseased tissues.

NovoGen MMX Bioprinter™

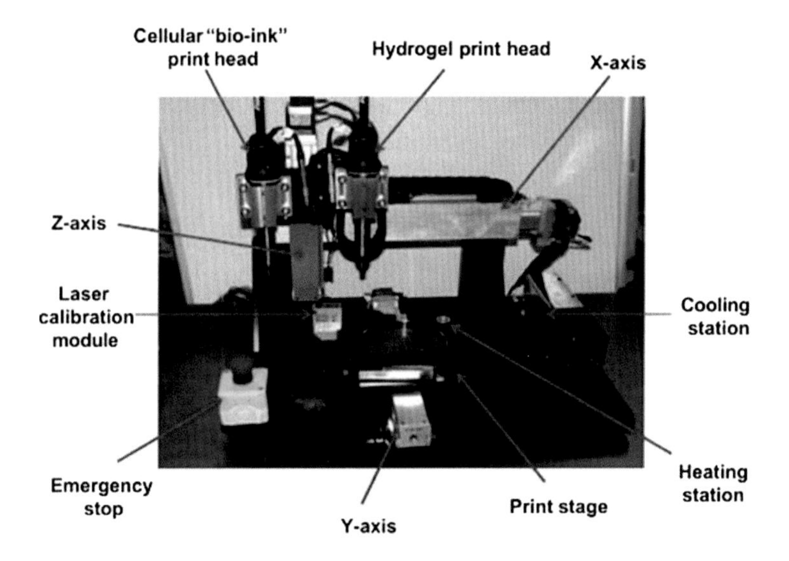

The printer includes two robotically controlled precision printheads: one for placing human cells and the other for placing a hydrogel, scaffold, or support matrix. One of the most complex challenges in the development of the bioprinter was perfecting a means of consistently positioning the cell dispensing capillary tip attached to the printhead within microns. Invetech developed a computer-controlled, laser-based calibration system to achieve the required repeatability.

While the science and engineering is incredibly complex, the intuitive user interface makes operating the printer extremely simple: the operator literally draws the organ to be built on the computer screen.

The initial focus of the bioprinter is on producing simple tissues, like blood vessels and nerve conduit, but potentially any tissue or organ can be built.

MICROEXTRUSION-BASED BIOPRINTING

Bioprinting methods based on extrusion of cell or cell-laden biomaterials (Table 4.2; Figs. 4.1 and 4.2) use self-assembly cells to construct 3D biological constructs. The main principle of extrusion-based bioprinting techniques is to force continuous filaments of materials including hydrogels, biocompatible copolymers, and living cells through a nozzle with the help of a computer to construct a 3D structure [20–27]. Extrusion-based printers usually have a temperature-controlled material handling and dispensing system and stage with movement capability along the x, y, and z axes. The printers are directed by computer-aided design and computer-aided manufacturing (CAD-CAM) software and continuous filaments are deposited in two dimensions layer-by-layer to from 3D tissue constructs. The stage or the extrusion head is moved along the

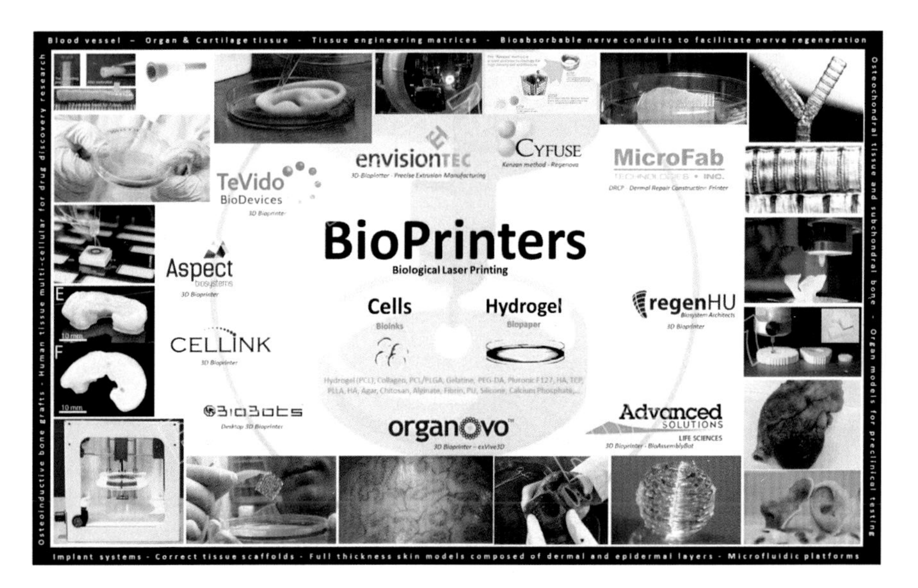

Figure 4.2 Current industrial, commercial research-use-only bioprinters.

z axis, and the printed layers serve as a base and support for the next layer. Pneumatic or mechanical (piston or screw) are the most common techniques to print biological materials for 3D bioprinting applications [32–69]. Additionally, novel multinozzle biopolymer deposition systems have been developed for freeform fabrication of biopolymer-based tissue scaffolds and cell-embedded tissue constructs [70,71].

Extrusion-based printing allows the construction of organized structures within a realistic timeframe, and hence it is the most promising bioprinting technology. The main advantage of extrusion-based bioprinting is the ability to print very high cell densities. Some groups have developed 3D bioprinters in order to use multicellular spheroids or cylinders as bioink to create 3D tissue constructs [17–19,21,25,78–89]. However, preparing bioink requires time-consuming manual operation and makes totally automated and computer-controlled 3D bioprinting impossible in earlier studies. Therefore our group has focused on developing a continuous bioprinting approach in order to extrude cylindrical multicellular aggregates using an extrusion-based bioprinter, which is an automated, flexible platform designed to fabricate 3D tissue-engineered cell constructs. In order to bioprint anatomically correct tissue constructs directly from medical images, the targeted tissue or organ must be biomodeled.

BIOBOTS

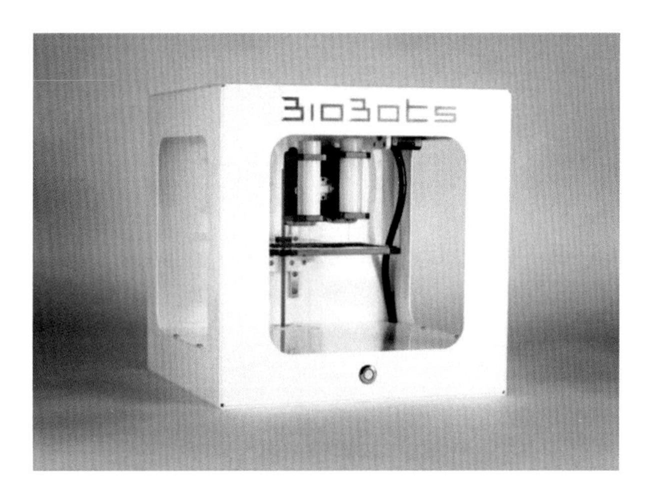

The vision of BioBots is to make tools that harness life as an engineering discipline and push the human race forward. BioBots was created in a dorm room on top of a noisy college bar with the mission of conquering the largest mystery of our generation—life. Disillusioned with existing tools and technologies for engineering organisms, and inspired by the idea of biology as technology, BioBots was launched with the motto "Build with Life."

It only took a few weeks for the first apostles to join us. Dr. Dan Huh and his student Yooni at Penn began working with a prototype that would become the first BioBot. With the help and unyielding support of early clients and partners such as Elliot Menschik at DreamIt Health, the company began the journey of bringing biofabrication technology to people across the world.

Today hundreds of labs are turning to BioBots for tools that allow them to engineer biology at lower cost.

REFERENCES

[1] Lang-8. Bioprinting: brief timeline and outlook, <http://lang8.com/1017887/journals/107384577998436484655804992513119667379>.

[2] Miller JS, Stevens KR, Yang MT, Baker BM, Nguyen D-HT, Cohen DM, et al. Rapid casting of patterned vascular networks for perfusable engineered three-dimensional tissues. Nat Mater 2012;11(9):768–74.

[3] Kolesky DB, Truby RL, Gladman AS, Busbee TA, Homan KA, Lewis JA. 3D bioprinting of vascularized, heterogeneous cell-laden tissue constructs. Adv Mater 2014;26(19):3124–30.

[4] Langer R, Vacanti JP. Tissue engineering. Science 1993;260(5110):920–6.

[5] Pollok JM, Vacanti JP. Tissue engineering. Semin Pediatr Surg 1996;5(3):191–6.

[6] Nakayama K. In vitro biofabrication of tissues and organs. In: Forgacs G, Sun W, editors. Biofabrication: micro- and nano-fabrication, printing, patterning and assemblies. Amsterdam: Elsevier; 2013. p. 1–16. Available from: http://dx.doi.org/10.1016/B978-1-4557-2852-7.00013-5.

[7] Atala A, Bauer SB, Soker S, Yoo JJ, Retik AB. Tissue-engineered autologous bladders for patients needing cystoplasty. Lancet 2006;367(9518):1241–6.

[8] Brittberg M, Nilsson A, Lindahl A, Ohlsson C, Peterson L. Rabbit articular cartilage defects treated with autologous cultured chondrocytes. Clin Orthop Relat Res 1996;326:270–83.

[9] Gallico GG, O'Connor NE, Compton CC, Kehinde O, Green H. Permanent cover- age of large burn wounds with autologous cultured human epithelium. N Engl J Med 1984;311 (7):448–51.

[10] Hibino N, McGillicuddy E, Matsumura G, Ichihara Y, Naito Y, Breuer C, et al. Late- term results of tissue-engineered vascular grafts in humans. J Thorac Cardiovasc Surg 2010;139 (2):431–6, e2.

[11] Matsumura G, Ishihara Y, Miyagawa-Tomita S, Ikada Y, Matsuda S, Kurosawa H, et al. Evaluation of tissue-engineered vascular autografts. Tissue Eng 2006;12(11):3075–83.

[12] Niemeyer P, Krause U, Fellenberg J, Kasten P, Seckinger A, Ho AD, et al. Evaluation of mineralized collagen and alphatricalcium phosphate as scaffolds for tissue engineering of bone using human mesenchymal stem cells. Cells Tissues Organs 2004;177(2):880–5.

[13] Oberpenning F, Meng J, Yoo JJ, Atala A. De novo reconstitution of a functional mammalian urinary bladder by tissue engineering. Nat Biotechnol 1999;17(2):149–55.

[14] Radisic M, Park H, Shing H, Consi T, Schoen FJ, Langer R, et al. Functional assembly of engineered myocardium by electrical stimulation of cardiac myocytes cultured on scaffolds. Proc Natl Acad Sci U S A 2004;101(52):18129–34.

[15] Shin'oka T, Matsumura G, Hibino N, Naito Y, Watanabe M, Konuma T, et al. Midterm clinical result of tissue-engineered vascular autografts seeded with autologous bone marrow cells. J Thorac Cardiovasc Surg 2005;129(6):1330–8.

[16] Nakamura M, Iwanaga S, Henmi C, Arai K, Nishiyama Y. Biomatrices and biomaterials for future developments of bioprinting and biofabrication. Biofabrication 2010;2(1):014110.

[17] Wüst S, Müller R, Hofmann S. Controlled positioning of cells in biomaterials—approaches towards 3D tissue printing. J Funct Biomater 2011;2(3):119–54.

[18] Melchels FPW, Domingos MAN, Klein TJ, Malda J, Bartolo PJ, Hutmacher DW. Additive manufacturing of tissues and organs. Prog Polym Sci 2012;37(8):1079–104.

[19] Khoda AKM, Ozbolat IT, Koc B. A functionally gradient variational porosity architecture for hollowed scaffolds fabrication. Biofabrication 2011;3(3):034106.

[20] Khoda AKM, Ozbolat IT, Koc B. Designing heterogeneous porous tissue scaffolds for additive manufacturing processes. Comput Aided Des 2013;45(12):1507–23.

[21] Matsuda N, Shimizu T, Yamato M, Okano T. Tissue engineering based on cell sheet technology. Adv Mater 2007;19(20):3089–99.

[22] Mironov V, Prestwich G, Forgacs G. Bioprinting living structures. J Mater Chem 2007;17(20):2054–60.

[23] Mironov V, Visconti RP, Kasyanov V, Forgacs G, Drake CJ, Markwald RR. Organ printing: tissue spheroids as building blocks. Biomaterials 2009;30(12):2164–74.

[24] Jakab K, Norotte C, Marga F, Murphy K, Vunjak-Novakovic G, Gabor F. Tissue engineering by self-assembly and bio-printing of living cells. Biofabrication 2010;2(2):022001.

[25] L'Heureux N, Dusserre N, Konig G, Victor B, Keire P, Wight TN, et al. Human tissue-engineered blood vessels for adult arterial revascularization. Nat Med 2006;12(3):361–5.

[26] L'Heureux N, Pâquet S, Labbé R, Germain L, Auger FA. A completely biological tissue-engineered human blood vessel. FASEB J 1998;12(1):47–56.

[27] Ferris C, Gilmore K, Wallace G, in het Panhuis M. Biofabrication: an overview of the approaches used for printing of living cells. Appl Microbiol Biotechnol 2013;97(10):4243–58.

[28] Marga F, Jakab K, Khatiwala C, Shepherd B, Dorfman S, Hubbard B, et al. Toward engineering functional organ modules by additive manufacturing. Biofabrication 2012;4 (2):022001.

[29] Forgacs G, Foty RA, Shafrir Y, Steinberg MS. Viscoelastic properties of living embryonic tissues: a quantitative study. Biophys J 1998;74(5):2227–34.

[30] Koç B, Hafezi F, Ozler SB, Kucukgul C. Bioprinting-application of additive manufacturing in medicine. In: Bandyopadhyay A, Bose S, editors. Additive manufacturing. Boca Raton, FL: CRC Press; 2016.

[31] Chang CC, Boland ED, Williams SK, Hoying JB. Direct-write bioprinting three-dimensional biohybrid systems for future regenerative therapies. J Biomed Mater Res B Appl Biomater 2011;98(1):160–70.

[32] Odde DJ, Renn MJ. Laser-guided direct writing for applications in biotechnology. Trends Biotechnol 1999;17(10):385–9.

[33] Nahmias Y, Schwartz RE, Verfaillie CM, Odde DJ. Laser-guided direct writing for three-dimensional tissue engineering. Biotechnol Bioeng 2005;92(2):129–36.

[34] Zhen M, Russell KP, Qin W, Julie XY, Xiaocong Y, Peng X, et al. Laser-guidance-based cell deposition microscope for heterotypic single-cell micropatterning. Biofabrication 2011;3(3):034107.

[35] Murphy SV, Atala A. 3D bioprinting of tissues and organs. Nat Biotechnol 2014; 32(8):773–85.

[36] Guillotin B, Souquet A, Catros S, Duocastella M, Pippenger B, Bellance S, et al. Laser assisted bioprinting of engineered tissue with high cell density and microscale organization. Biomaterials 2010;31(28):7250–6.

[37] Barron JA, Ringeisen BR, Kim H, Spargo BJ, Chrisey DB. Application of laser printing to mammalian cells. Thin Solid Films 2004;453–454:383–7.

[38] Ringeisen BR, Kim H, Barron JA, Krizman DB, Chrisey DB, Jackman S, et al. Laser printing of pluripotent embryonal carcinoma cells. Tissue Eng 2004;10(3–4):483–91.

[39] Guillemot F, Souquet A, Catros S, Guillotin B, Lopez J, Faucon M, et al. High-throughput laser printing of cells and biomaterials for tissue engineering. Acta Biomater 2010;6(7):2494–500.

[40] Haraguchi Y, Shimizu T, Yamato M, Okano T. Regenerative therapies using cell sheet-based tissue engineering for cardiac disease. Cardiol Res Pract 2011;2011:845170.

[41] Shimizu T, Yamato M, Isoi Y, Akutsu T, Setomaru T, Abe K, et al. Fabrication of pulsatile cardiac tissue grafts using a novel 3-dimensional cell sheet manipulation technique and temperature-responsive cell culture surfaces. Circ Res 2002;90(3):e40.

[42] Imen Elloumi H, Masayuki Y, Teruo O. Cell sheet technology and cell patterning for biofabrication. Biofabrication 2009;1(2):022002.

[43] Shimizu T, Yamato M, Akutsu T, Shibata T, Isoi Y, Kikuchi A, et al. Electrically communicating three-dimensional cardiac tissue mimic fabricated by layered cultured cardiomyocyte sheets. J Biomed Mater Res 2002;60(1):110–17.

[44] Haraguchi Y, Shimizu T, Yamato M, Kikuchi A, Okano T. Electrical coupling of cardio- myocyte sheets occurs rapidly via functional gap junction formation. Biomaterials 2006;27(27):4765–74.

[45] Shimizu T, Sekine H, Isoi Y, Yamato M, Kikuchi A, Okano T. Long-term survival and growth of pulsatile myocardial tissue grafts engineered by the layering of cardiomyocyte sheets. Tissue Eng 2006;12(3):499–507.

[46] Yamato M, Utsumi M, Kushida A, Konno C, Kikuchi A, Okano T. Thermo-responsive culture dishes allow the intact harvest of multilayered keratinocyte sheets without dispase by reducing temperature. Tissue Eng 2001;7(4):473–80.

[47] Kushida A, Yamato M, Isoi Y, Kikuchi A, Okano T. A noninvasive transfer system for polarized renal tubule epithelial cell sheets using temperature-responsive culture dishes. Eur Cell Mater 2005;10:23–30.

[48] Akizuki T, Oda S, Komaki M, Tsuchioka H, Kawakatsu N, Kikuchi A, et al. Application of periodontal ligament cell sheet for periodontal regeneration: a pilot study in beagle dogs. J Periodontal Res 2005;40(3):245–51.

[49] Hasegawa M, Yamato M, Kikuchi A, Okano T, Ishikawa I. Human periodontal ligament cell sheets can regenerate periodontal ligament tissue in an athymic rat model. Tissue Eng 2005;11(3–4):469–78.

[50] Malda J, Visser J, Melchels FP, Jüngst T, Hennink WE, Dhert WJA, et al. 25th anniversary article: engineering hydrogels for biofabrication. Adv Mater 2013;25(36):5011–28.

[51] Nakamura M, Kobayashi A, Takagi F, Watanabe A, Hiruma Y, Ohuchi K, et al. Biocompatible inkjet printing technique for designed seeding of individual living cells. Tissue Eng 2005;11(11–12):1658–66.

[52] Wilson WC, Boland T. Cell and organ printing 1: protein and cell printers. Anat Rec A Discov Mol Cell Evol Biol 2003;272(2):491–6.

[53] Boland T, Mironov V, Gutowska A, Roth EA, Markwald RR. Cell and organ printing 2: fusion of cell aggregates in three-dimensional gels. Anat Rec A Discov Mol Cell Evol Biol 2003;272(2):497–502.

[54] Xu T, Jin J, Gregory C, Hickman JJ, Boland T. Inkjet printing of viable mammalian cells. Biomaterials 2005;26(1):93–9.

[55] Cui X, Boland T. Human microvasculature fabrication using thermal inkjet printing technology. Biomaterials 2009;30(31):6221–7.

[56] Boland T, Tao X, Damon BJ, Manley B, Kesari P, Jalota S, et al. Drop-on-demand printing of cells and materials for designer tissue constructs. Mater Sci Eng C 2007;27(3):372–6.

[57] Nakamura M, Nishiyama Y, Henmi C, Yamaguchi K, Mochizuki S, Koki T, et al. Inkjet bio- printing as an effective tool for tissue fabrication. NIP Digital Fabr Conf 2006;2006 (3):89–92.

[58] Nishiyama Y, Nakamura M, Henmi C, Yamaguchi K, Mochizuki S, Nakagawa H, et al. Development of a three-dimensional bioprinter: construction of cell supporting structures using hydrogel and state-of-the-art inkjet technology. J Biomech Eng 2008;131(3):035001.

[59] Xu C, Chai W, Huang Y, Markwald RR. Scaffold-free inkjet printing of three-dimensional zigzag cellular tubes. Biotechnol Bioeng 2012;109(12):3152–60.

[60] Khatiwala C, Law R, Shepherd B, Dorfman S, Csete M. 3D cell bioprinting for regenerative medicine research and therapies. Gene Ther Regul 2012;07(01):1230004.

[61] Shabnam P, Madhuja G, Frédéric L, Karen CC. Effects of surfactant and gentle agitation on inkjet dispensing of living cells. Biofabrication 2010;2(2):025003.

[62] Chahal D, Ahmadi A, Cheung KC. Improving piezoelectric cell printing accuracy and reliability through neutral buoyancy of suspensions. Biotechnol Bioeng 2012;109(11):2932–40.

[63] Ferris CJ, Gilmore KJ, Beirne S, McCallum D, Wallace GG, in het Panhuis M. Bio-ink for on-demand printing of living cells. Biomater Sci 2013;1(2):224–30.

[64] Foty RA, Steinberg MS. The differential adhesion hypothesis: a direct evaluation. Dev Biol 2005;278(1):255–63.

[65] Steinberg MS. Differential adhesion in morphogenesis: a modern view. Curr Opin Genet Dev 2007;17(4):281–6.

[66] Steinberg MS. Reconstruction of tissues by dissociated cells. Some morphogenetic tissue movements and the sorting out of embryonic cells may have a common explanation. Science 1963;141(3579):401–8.

[67] Jakab K, Norotte C, Damon B, Marga F, Neagu A, Besch-Williford CL, et al. Tissue engineering by self-assembly of cells printed into topologically defined structures. Tissue Eng 2008;14(3):413–21.

[68] Kelm JM, Lorber V, Snedeker JG, Schmidt D, Broggini-Tenzer A, Weisstanner M, et al. A novel concept for scaffold-free vessel tissue engineering: self-assembly of microtissue building blocks. J Biotechnol 2010;148(1):46–55.

[69] Gwyther TA, Hu JZ, Christakis AG, Skorinko JK, Shaw SM, Billiar KL, et al. Engineered vascular tissue fabricated from aggregated smooth muscle cells. Cells Tissues Organs 2011;194(1):13–24.

[70] Adebayo O, Hookway TA, Hu JZ, Billiar KL, Rolle MW. Self-assembled smooth muscle cell tissue rings exhibit greater tensile strength than cell-seeded fibrin or collagen gel rings. J Biomed Mater Res A 2013;101(2):428–37.

[71] Napolitano AP, Chai P, Dean DM, Morgan JR. Dynamics of the self-assembly of complex cellular aggregates on micromolded nonadhesive hydrogels. Tissue Eng 2007;13(8):2087–94.

[72] Dean DM, Napolitano AP, Youssef J, Morgan JR. Rods, tori, and honeycombs: the directed self-assembly of microtissues with prescribed microscale geometries. FASEB J 2007; 21(14):4005–12.

[73] Khalil S, Nam J, Sun W. Multi-nozzle deposition for construction of 3D biopolymer tissue scaffolds. Rapid Prototyp J 2005;11(1):9–17.

[74] Khalil S, Sun W. Biopolymer deposition for freeform fabrication of hydrogel tissue constructs. Mater Sci Eng C 2007;27(3):469–78.

[75] Smith CM, Christian JJ, Warren WL, Williams SK. Characterizing environmental factors that impact the viability of tissue-engineered constructs fabricated by a direct-write bioassembly tool. Tissue Eng 2007;13(2):373–83.

[76] Smith CM, Stone AL, Parkhill RL, Stewart RL, Simpkins MW, Kachurin AM, et al. Three-dimensional bioassembly tool for generating viable tissue-engineered constructs. Tissue Eng 2004;10(9–10):1566–76.

[77] Duan B, Hockaday LA, Kang KH, Butcher JT. 3D Bioprinting of heterogeneous aortic valve conduits with alginate/gelatin hydrogels. J Biomed Mater Res A 2013;101(5):1255–64.

[78] Hockaday LA, Kang KH, Colangelo NW, Cheung PYC, Duan B, Malone E, et al. Rapid 3D printing of anatomically accurate and mechanically heterogeneous aortic valve hydrogel scaffolds. Biofabrication 2012;4(3):035005.

[79] Chang R, Nam J, Sun W. Effects of dispensing pressure and nozzle diameter on cell survival from solid freeform fabrication-based direct cell writing. Tissue Eng Part A 2008;14(1):41–8.

[80] Cohen DL, Malone E, Lipson H, Bonassar LJ. Direct freeform fabrication of seeded hydrogels in arbitrary geometries. Tissue Eng 2006;12(5):1325–35.

[81] Christopher MO, Francoise M, Gabor F, Cheryl MH. Biofabrication and testing of a fully cellular nerve graft. Biofabrication 2013;5(4):045007.

[82] Mironov V, Kasyanov V, Markwald RR. Organ printing: from bioprinter to organ biofabrication line. Curr Opin Biotechnol 2011;22(5):667–73.

[83] Norotte C, Marga FS, Niklason LE, Forgacs G. Scaffold-free vascular tissue engineering using bioprinting. Biomaterials 2009;30(30):5910–17.

[84] Kucukgul C, Ozler SB, Inci I, Karakas E, Irmak S, Gozuacik D, et al. 3D bioprinting of biomimetic aortic vascular constructs with self-supporting cells. Biotechnol Bioeng 2014;112(4):811–21.

[85] Ozler SB, Kucukgul C, Koc B. 3D continuous cell bioprinting. submitted October 1, 2016.

[86] Kucukgul C, Ozler B, Karakas HE, Gozuacik D, Koc B. 3D hybrid bioprinting of macrovascular structures. Procedia Eng 2013;59:183–92.

[87] Peltola SM, Melchels FP, Grijpma DW, Kellomaki M. A review of rapid prototyping techniques for tissue engineering purposes. Ann Med 2008;40:268–80.

[88] Hutmacher DW, Sittinger M, Risbud MV. Scaffold-based tissue engineering: rationale for computer-aided design and solid free-form fabrication systems. Trends Biotechnol 2004;22:354–62.

[89] Klebe RJ. Cytoscribing: a method for micropositioning cells and the construction of two- and three-dimensional synthetic tissues. Exp Cell Res 1988;179:362–73.

[90] Xu T, et al. Complex heterogeneous tissue constructs containing multiple cell types prepared by inkjet printing technology. Biomaterials 2013;34:130–9.

[91] Cui X, Boland T, D'Lima DD, Lotz MK. Thermal inkjet printing in tissue engineering and regenerative medicine. Recent Pat Drug Deliv Formul 2012;6:149–55.

[92] Iwami K, et al. Bio rapid prototyping by extruding/aspirating/refilling thermoreversible hydrogel. Biofabrication 2010;2:014108.

[93] Shor L, et al. Precision extruding deposition (PED) fabrication of polycaprolactone (PCL) scaffolds for bone tissue engineering. Biofabrication 2009;1:015003.

[94] Barron JA, Wu P, Ladouceur HD, Ringeisen BR. Biological laser printing: a novel technique for creating heterogeneous 3-dimensional cell patterns. Biomed Microdevices 2004;6:139–47.

[95] Xu T, Kincaid H, Atala A, Yoo JJ. High-throughput production of single-cell microparticles using an inkjet printing technology. J Manuf Sci Eng 2008;130(2):021017.

[96] Xu T, et al. Characterization of cell constructs generated with inkjet printing technology using in vivo magnetic resonance imaging. J Manuf Sci Eng 2008;130(2):021013.

[97] Okamoto T, Suzuki T, Yamamoto N. Microarray fabrication with covalent attachment of DNA using bubble jet technology. Nat Biotechnol 2000;18:438–41.

[98] Goldmann T, Gonzalez JS. DNA-printing: utilization of a standard inkjet printer for the transfer of nucleic acids to solid supports. J Biochem Biophys Methods 2000;42:105–10.

[99] Xu T, et al. Viability and electrophysiology of neural cell structures generated by the inkjet printing method. Biomaterials 2006;27:3580–8.

[100] Cui X, Dean D, Ruggeri ZM, Boland T. Cell damage evaluation of thermal inkjet printed Chinese hamster ovary cells. Biotechnol Bioeng 2010;106:963–9.

[101] Tekin E, Smith PJ, Schubert US. Inkjet printing as a deposition and patterning tool for polymers and inorganic particles. Soft Matter 2008;4:703–13.

[102] Fang Y, et al. Rapid generation of multiplexed cell cocultures using acoustic droplet ejection followed by aqueous two-phase exclusion patterning. Tissue Eng Part C Methods 2012;18:647–57.

[103] Demirci U, Montesano G. Single cell epitaxy by acoustic picolitre droplets. Lab Chip 2007;7:1139–45.

[104] Saunders R, Bosworth L, Gough J, Derby B, Reis N. Selective cell delivery for 3D tissue culture and engineering. Eur Cell Mater 2004;7:84–5.

[105] Saunders RE, Gough JE, Derby B. Delivery of human fibroblast cells by piezoelectric drop-on-demand inkjet printing. Biomaterials 2008;29:193–203.

[106] Tasoglu S, Demirci U. Bioprinting for stem cell research. Trends Biotechnol 2013;31:10–19.

[107] Kim JD, Choi JS, Kim BS, Chan Choi Y, Cho YW. Piezoelectric inkjet printing of polymers: stem cell patterning on polymer substrates. Polymer (Guildf) 2010;51:2147–54.

[108] Murphy SV, Skardal A, Atala A. Evaluation of hydrogels for bio-printing applications. J Biomed Mater Res A 2013;101:272–84.

[109] Hennink WE, van Nostrum CF. Novel crosslinking methods to design hydrogels. Adv Drug Deliv Rev 2002;54:13–36.

[110] Skardal A, Zhang J, Prestwich GD. Bioprinting vessel-like constructs using hyaluronan hydrogels crosslinked with tetrahedral polyethylene glycol tetracrylates. Biomaterials 2010;31:6173–81.

[111] Campbell PG, Miller ED, Fisher GW, Walker LM, Weiss LE. Engineered spatial patterns of FGF-2 immobilized on fibrin direct cell organization. Biomaterials 2005;26:6762–70.

[112] Phillippi JA, et al. Microenvironments engineered by inkjet bioprinting spatially direct adult stem cells toward muscle- and bone-like subpopulations. Stem Cells 2008;26:127–34.

[113] Sekitani T, Noguchi Y, Zschieschang U, Klauk H, Someya T. Organic transistors manufactured using inkjet technology with subfemtoliter accuracy. Proc Natl Acad Sci USA 2008;105:4976–80.

[114] Singh M, Haverinen HM, Dhagat P, Jabbour GE. Inkjet printing-process and its applications. Adv Mater 2010;22:673–85.

[115] Skardal A, et al. Bioprinted amniotic fluid-derived stem cells accelerate healing of large skin wounds. Stem Cells Transl Med 2012;1:792–802.

[116] Cui X, Breitenkamp K, Finn MG, Lotz M, D'Lima DD. Direct human cartilage repair using three-dimensional bioprinting technology. Tissue Eng Part A 2012;18:1304–12.

[117] Xu T, et al. Hybrid printing of mechanically and biologically improved constructs for cartilage tissue engineering applications. Biofabrication 2013;5:015001.

[118] De Coppi P, et al. Isolation of amniotic stem cell lines with potential for therapy. Nat Biotechnol 2007;25:100–6.

[119] Jones N. Science in three dimensions: the print revolution. Nature 2012;487:22–3.

[120] Fedorovich NE, et al. Evaluation of photocrosslinked Lutrol hydrogel for tissue printing applications. Biomacromolecules 2009;10:1689–96.

[121] Jakab K, Damon B, Neagu A, Kachurin A, Forgacs G. Three-dimensional tissue constructs built by bioprinting. Biorheology 2006;43:509–13.

[122] Visser J, et al. Biofabrication of multi-material anatomically shaped tissue constructs. Biofabrication 2013;5:035007.

[123] Censi R, et al. The tissue response to photopolymerized PEG-p(HPMAm-lactate)-based hydrogels. J Biomed Mater Res A 2011;97:219–29.

[124] Schuurman W, et al. Gelatin-methacrylamide hydrogels as potential biomaterials for fabrication of tissue-engineered cartilage constructs. Macromol Biosci 2013;13:551–61.

[125] Guvendiren M, Lu HD, Burdick JA. Shear-thinning hydrogels for biomedical applications. Soft Matter 2012;8:260–72.

[126] Marga, F., et al. Organ printing: a novel tissue engineering paradigm. In: 5th European Conference of the International Federation for Medical and Biological Engineering 14–18 September 2011, Budapest, Hungary. Springer; 2012. p. 27–30.

[127] Nair K, et al. Characterization of cell viability during bioprinting processes. Biotechnol J 2009;4:1168–77.

[128] Skardal A, Zhang J, McCoard L, Oottamasathien S, Prestwich GD. Dynamically crosslinked gold nanoparticle—hyaluronan hydrogels. Adv Mater 2010;22:4736–40.

[129] Skardal A, et al. Photocrosslinkable hyaluronan-gelatin hydrogels for two-step bioprinting. Tissue Eng Part A 2010;16:2675–85.

[130] Chang R, Nam J, Sun W. Direct cell writing of 3D microorgan for *in vitro* pharmacokinetic model. Tissue Eng Part C Methods 2008;14:157–66.

[131] Xu F, et al. A three-dimensional *in vitro* ovarian cancer coculture model using a high-throughput cell patterning platform. Biotechnol J 2011;6:204–12.

[132] Bohandy J, Kim B, Adrian F. Metal deposition from a supported metal film using an excimer laser. J Appl Phys 1986;60:1538–9.

[133] Chrisey DB. Materials processing: the power of direct writing. Science 2000;289:879–81.

[134] Colina M, Serra P, Fernandez-Pradas JM, Sevilla L, Morenza JL. DNA deposition through laser induced forward transfer. Biosens Bioelectron 2005;20:1638–42.

[135] Dinca V, et al. Directed three-dimensional patterning of self-assembled peptide fibrils. Nano Lett 2008;8:538–43.

[136] Guillemot F, Souquet A, Catros S, Guillotin B. Laser-assisted cell printing: principle, physical parameters versus cell fate and perspectives in tissue engineering. Nanomedicine 2010;5:507–15.

[137] Hopp B, et al. Survival and proliferative ability of various living cell types after laser-induced forward transfer. Tissue Eng 2005;11:1817–23.

[138] Gruene M, et al. Laser printing of stem cells for biofabrication of scaffold-free autologous grafts. Tissue Eng Part C Methods 2011;17:79–87.

[139] Koch L, et al. Laser printing of skin cells and human stem cells. Tissue Eng Part C Methods 2010;16:847–54.

[140] Guillotin B, Guillemot F. Cell patterning technologies for organotypic tissue fabrication. Trends Biotechnol 2011;29:183–90.

[141] Kattamis NT, Purnick PE, Weiss R, Arnold CB. Thick film laser induced forward transfer for deposition of thermally and mechanically sensitive materials. Appl Phys Lett 2007;91:171120–3.

[142] Duocastella M, Fernandez-Pradas J, Morenza J, Zafra D, Serra P. Novel laser printing technique for miniaturized biosensors preparation. Sens Actuators B Chem 2010; 145:596–600.

[143] Michael S, et al. Tissue engineered skin substitutes created by laser-assisted bioprinting form skin-like structures in the dorsal skin fold chamber in mice. PLoS One 2013;8:e57741.

[144] Keriquel V, et al. *In vivo* bioprinting for computer- and robotic-assisted medical intervention: preliminary study in mice. Biofabrication 2010;2:014101.

[145] Hunt NC, Grover LM. Cell encapsulation using biopolymer gels for regenerative medicine. Biotechnol Lett 2010;32:733–42.

[146] Sun J, et al. Chitosan functionalized ionic liquid as a recyclable biopolymer-supported catalyst for cycloaddition of CO2. Green Chem 2012;14:654–60.

[147] Spiller KL, Maher SA, Lowman AM. Hydrogels for the repair of articular cartilage defects. Tissue Eng Part B Rev 2011;17:281–99.

[148] Li Z, Kawashita M. Current progress in inorganic artificial biomaterials. J Artif Organs 2011;14:163–70.

[149] Talbot EL, Berson A, Brown PS, Bain CD. Evaporation of picoliter droplets on surfaces with a range of wettabilities and thermal conductivities. Phys Rev E 2012;85:061604.

[150] Hopp BL, et al. Femtosecond laser printing of living cells using absorbing film-assisted laser-induced forward transfer. Opt Eng 2012;51:014302–6.

[151] Williams DF. On the mechanisms of biocompatibility. Biomaterials 2008;29:2941–53.

[152] West JL, Hubbell JA. Polymeric biomaterials with degradation sites for proteases involved in cell migration. Macromolecules 1999;32:241–4.

[153] Ananthanarayanan A, Narmada BC, Mo X, McMillian M, Yu H. Purpose-driven biomaterials research in liver-tissue engineering. Trends Biotechnol 2011;29:110–18.
Hutmacher DW. Scaffolds in tissue engineering bone and cartilage. Biomaterials 2000;21:2529–43.

[154] Limpanuphap S, Derby B. Manufacture of biomaterials by a novel printing process. J Mater Sci Mater Med 2002;13:1163–6.

[155] Zhang S, et al. Self-complementary oligopeptide matrices support mammalian cell attachment. Biomaterials 1995;16:1385–93.

[156] Hersel U, Dahmen C, Kessler H. RGD modified polymers: biomaterials for stimulated cell adhesion and beyond. Biomaterials 2003;24:4385–415.

[157] Karp JM, et al. Controlling size, shape and homogeneity of embryoid bodies using poly (ethylene glycol) microwells. Lab Chip 2007;7:786–94.

[158] Teixeira AI, Nealey PF, Murphy CJ. Responses of human keratocytes to micro- and nanostructured substrates. J Biomed Mater Res A 2004;71:369–76.

[159] Price RL, Haberstroh KM, Webster TJ. Enhanced functions of osteoblasts on nano-structured surfaces of carbon and alumina. Med Biol Eng Comput 2003;41:372–5.

[160] Behonick DJ, Werb Z. A bit of give and take: the relationship between the extracellular matrix and the developing chondrocyte. Mech Dev 2003;120:1327–36.

[161] Discher DE, Janmey P, Wang YL. Tissue cells feel and respond to the stiffness of their substrate. Science 2005;310:1139–43.

[162] Stevens MM, George JH. Exploring and engineering the cell surface interface. Science 2005;310:1135–8.

[163] Baptista PM, et al. Whole organ decellularization–a tool for bioscaffold fabrication and organ bioengineering. Conf Proc IEEE Eng Med Biol Soc 2009;2009:6526–9.

[164] Sullivan DC, et al. Decellularization methods of porcine kidneys for whole organ engineering using a high-throughput system. Biomaterials 2012;33:7756–64.

[165] Hynes RO, Naba A. Overview of the matrisome–an inventory of extracellular matrix constituents and functions. Cold Spring Harb Perspect Biol 2012;4:a004903.

[166] Ambesi-Impiombato FS, Parks LA, Coon HG. Culture of hormone-dependent functional epithelial cells from rat thyroids. Proc Natl Acad Sci U S A 1980;77:3455–9.

[167] Hamm A, Krott N, Breibach I, Blindt R, Bosserhoff AK. Efficient transfection method for primary cells. Tissue Eng 2002;8:235–45.

[168] Okumura N, et al. Enhancement on primate corneal endothelial cell survival *in vitro* by a ROCK inhibitor. Invest Ophthalmol Vis Sci 2009;50:3680–7.

[169] Yu Z, et al. ROCK inhibition with Y27632 promotes the proliferation and cell cycle progression of cultured astrocyte from spinal cord. Neurochem Int 2012;61:1114–20.

[170] Dimri GP, et al. A biomarker that identifies senescent human cells in culture and in aging skin *in vivo*. Proc Natl Acad Sci U S A 1995;92:9363–7.

[171] Reubinoff BE, Pera MF, Fong CY, Trounson A, Bongso A. Embryonic stem cell lines from human blastocysts: somatic differentiation *in vitro*. Nat Biotechnol 2000;18:399–404.

[172] Friedenstein AJ, et al. Precursors for fibroblasts in different populations of hematopoietic cells as detected by the *in vitro* colony assay method. Exp Hematol 1974;2:83–92.

[173] Dominici M, et al. Minimal criteria for defining multipotent mesenchymal stromal cells. The International Society for Cellular Therapy position statement. Cytotherapy 2006; 8:315–17.

[174] Pittenger MF, et al. Multilineage potential of adult human mesenchymal stem cells. Science 1999;284:143–7.

[175] Zuk PA, et al. Human adipose tissue is a source of multipotent stem cells. Mol Biol Cell 2002;13:4279–95.

[176] Murphy S, et al. Amnion epithelial cell isolation and characterization for clinical use. Curr Protoc Stem Cell Biol 2010;1E6. Available from: http://dx.doi.org/10.1002/9780470151808.sc01e06s13.

[177] Gillette BM, Jensen JA, Wang M, Tchao J, Sia SK. Dynamic hydrogels: switching of 3D microenvironments using two-component naturally derived extracellular matrices. Adv Mater 2010;22:686–91.

[178] Ott HC, et al. Perfusion-decellularized matrix: using nature's platform to engineer a bioartificial heart. Nat Med 2008;14:213–21.

[179] Chun SY, et al. Identification and characterization of bioactive factors in bladder submucosa matrix. Biomaterials 2007;28:4251–6.

[180] Schuurman W, et al. Bioprinting of hybrid tissue constructs with tailorable mechanical properties. Biofabrication 2011;3:021001.

[181] Ding S, et al. Synthetic small molecules that control stem cell fate. Proc Natl Acad Sci USA 2003;100:7632–7.

[182] Li XJ, et al. Directed differentiation of ventral spinal progenitors and motor neurons from human embryonic stem cells by small molecules. Stem Cells 2008;26:886–93.

[183] Chen S, et al. A small molecule that directs differentiation of human ESCs into the pancreatic lineage. Nat Chem Biol 2009;5:258–65.

[184] Chen S, Zhang Q, Wu X, Schultz PG, Ding S. Dedifferentiation of lineage-committed cells by a small molecule. J Am Chem Soc 2004;126:410–11.

[185] Visconti RP, et al. Towards organ printing: engineering an intra-organ branched vascular tree. Expert Opin Biol Ther 2010;10:409–20.

[186] Perez-Pomares JM, et al. *In vitro* self-assembly of proepicardial cell aggregates: an embryonic vasculogenic model for vascular tissue engineering. Anat Rec A Discov Mol Cell Evol Biol 2006;288:700–13.

[187] Tan Q, et al. Accelerated angiogenesis by continuous medium flow with vascular endothelial growth factor inside tissue-engineered trachea. Eur J Cardiothorac Surg 2007;31:806–11.

[188] Harrison BS, Eberli D, Lee SJ, Atala A, Yoo JJ. Oxygen producing biomaterials for tissue regeneration. Biomaterials 2007;28:4628–34.

[189] Salehi-Nik N, et al. Engineering parameters in bioreactor's design: a critical aspect in tissue engineering. Biomed Res Int 2013;2013:762132.

CHAPTER 5

Materials for Use in Bioprinting

Bioprinting is an extension of tissue engineering, where the cells are stacked in layers similar to the plastic/exotic filament 3D printing technique.

This chapter focuses on scaffold-free tissue engineering and its adaptation to the technology of 3D bioprinting. The challenges associated with 3D bioprinting using stem cells will also be discussed in this chapter.

Mammalian cells are not the only "bioink" utilized in bioprinting. Biodegradable or biocompatible materials are mostly used to build body parts or repair damaged ones as part of bioprinting. Some of the materials include certain types of flexible plastic, like the absorbable material used to make 3D printed sutures in a C-section, and titanium powder, which was used to create a hip implant. Using a NovoGen MMX bioprinter created by Organovo, the cells are layered between water-based layers until the tissue is built. The hydrogel in between layers is sometimes used to fill spaces in the tissue or as supports to the 3D printed tissue. Collagen is another material used to fuse cells together. This layer-by-layer approach is very similar to the normal 3D printing process, where products are built from the ground up.

Stem cells are also used in bioprinting. They adapt easily to tissues, and thus are an attractive option for bioprinting different organs and bones. Researchers at the University of Nottingham in the United Kingdom experimented with building bone replacements coated with stem cells that develop into tissues over time. The researchers said development of stem cell repair for complex tissues, like those that make up the heart or the liver.

BIOINK MATERIALS

The selection of bioink material is an important consideration when planning to bioprint a tissue model (Fig. 5.1). Materials must be suitable for crosslinking in order to facilitate bioprinter deposition and

Bioprinting. DOI: http://dx.doi.org/10.1016/B978-0-12-805369-0.00005-5

The selection of appropriate materials for use in bioprinting and their performance in a particular application depend on several features. These are listed below.

- Printability
  Properties that facilitate handling and deposition by the bioprinter may include viscosity, gelation methods and rheological properties.
- Biocompatibility
  Materials should not induce undesirable local or systemic responses from the host and should contribute actively and controllably to the biological and functional components of the construct.
- Degradation kinetics and byproducts
  Degradation rates should be matched to the ability of the cells to produce their own ECM; degradation byproducts should be nontoxic; materials should demonstrate suitable swelling or contractile characteristics.
- Structural and mechanical properties
  Materials should be chosen based on the required mechanical properties of the construct, ranging from rigid thermoplastic polymer fibers for strength to soft hydrogels for cell compatibility.
- Material biomimicry
  Engineering of desired structural, functional and dynamic material properties should be based on knowledge of tissue-specific endogenous material compositions.

Figure 5.1 Constituents considered highly favorable for bioprinting and implantation [1–44].

shape retention, as well as biocompatible for long-term transplantation goals. Table 5.1 outlines each of the key considerations in preparing an ideal bioink. There are countless possible biomaterials suitable for bioprinting including poly(ethylene glycol) (PEG), polylactic acid (PLA), polyglycolic acid (PGA), poly(lactic-*co*-glycolic) acid (PLGA), and gellan gum. The choice of material depends on the desired cell type and mechanical properties of the tissue being mimicked.

3D PRINTING BIODEGRADABLE POLYMERS

Some biodegradable polymers have been previously utilized for 3D printing in the traditional "melt" state. These include poly(caprolactone) (PCL), poly(L-lactide) (PLLA) and PLGA. The use of PLLA and PLGA for 3D printing of biodegradable porous scaffolds generated as potential bone graft materials has been reported. In this study 3D printing was used for fabrication rather than traditional techniques such as salt leaching/phase separation so as to allow the study of the effects of architecture and design on bone formation. Both PLLA and PLGA were printed using image-based design and indirect solid free-form fabrication. These structures were then seeded with BMP7-transduced gingival fibroblasts and implanted subcutaneously in mice for 4–8 weeks. Molecular computerised tomography (MCT) scans and histology revealed that the PLGA scaffolds had broken down after 4 weeks but the PLLA scaffolds (Fig. 5.2) maintained their architecture, which improved bone ingrowth revealing the importance of choosing the right material for the scaffold usage (see Saito E, Liao EE, Hu WW, Krebsbach PH, Hollister SJ.

Table 5.1 Matrix Bioinks Commonly Used for Bioprinting

Matrix Bioink	Source	Gelation Time	Gelation Process	Support/Sacrificial Material Needed?	References
Poly (ethylene glycol) diacrylate (PEGDA)	Synthetic	Minutes	Chemical	No	[4–8]
Methacrylated Chondroitin Sulfate	Natural	Minutes	Chemical	No	[16–19]
Alginate	Natural	Seconds	Chemical	No	[6–14]
Cellulose (various modifications)	Natural	Minutes	Chemical	No	[11–18]
Gellan Gum	Natural	Seconds	Chemical	No	[19–21]
Fibrin	Natural	Seconds	Chemical	Yes	[21–23,45,46]
Cell- and Tissue-derived ECM (Matrigel)	Natural	Minutes	Thermal	No	[47–50]
Collagen	Natural	0.5–1 h	Thermal	Yes	[50–53]
Spider Silk	Natural	Minutes	Thermal	No	[52,54]
Hyaluronic Acid (various modifications)	Natural	Modification Dependent	Modification Dependent	Modification Dependent	[6–9,24–44,51, 53,55–67]
Dextran (various modifications)	Natural	Modification Dependent	Modification Dependent	Modification Dependent	[56–59]
Gelatin (various modifications)	Natural	Modification Dependent	Modification Dependent	Modification Dependent	[45,46,53,60–63]

These bioinks can be used individually or combined to create hybrid bioinks. Some bioinks such as hyaluronic acid, dextran, and gelatin can be modified for improved gelation time and printability [45–67].

Figure 5.2 Example 3D-printed scaffold (bar is 1 mm). From: Seyednejad H, Gawlitta D, Dhert WJA, van Nostrum CF, Vermonden T, Hennink WE. Preparation and characterization of a 3D-printed scaffold based on a functionalized polyester for bone tissue engineering application. *Functional Polyes* 2012;7(5):87.

Effects of designed PLLA and 50: 50 PLGA scaffold architectures on bone formation in vivo. J Tissue Eng Regen Med 2013;7(2):99–11).

The parameters for 3D printing with polycaprolactone have recently been optimized. In this research synthesized PCL with a molecular weight near 79 kDa (similar to PolyVivo product AP11) was printed successfully using a Bioscaffolder (Envisiontec GmbH, Gladbeck, Germany) by extruding the molten polymer through a 23 Ga heated nozzle at 110°C and strands were applied onto a collector plate in a layer-by-layer method at a speed of 350 mm/minutes. The strand pattern was rotated at 90 degrees angles in between layers which created square pores. The distance between strands was 0.9 mm with a spindle speed of 200 rpm. The resultant elastic modulus of PCL was ~59 MPa, but the modulus (compressive) of the scaffold was around 10 MPa. Note the density of crystalline and the amorphous PCL was 1.200 and 1.021 g/cm^3, respectively (see Seyednejad H, Gawlitta D, Dhert WJA, van Nostrum CF, Vermonden T, Hennink WE. Preparation and characterization of a 3D-printed scaffold based on a functionalized polyester for bone tissue engineering application. *Functional Polyes* 2012;7(5):87).

THERMOGELS/GEL FORMERS

There are two types of temperature-responsive hydrogels. One is a "normal" type of gelation response in which the hydrogel-water solution melts to a liquid upon heating due to a reduction of chain-to-chain entanglements and an improvement in overall polymer solubility. Gelatin behaves this way and is a solid at cool temperatures but a liquid at higher temperatures (see Acharya G, Shin CS, McDermott M, Mishra H, Park H, Kwon IC, Park K. The hydrogel template method for fabrication of homogeneous nano/microparticles. J Controlled Release 2010;141(3):314–319). There is also a "reverse" gel response in which the hydrogel system transitions toward a solid phase upon heating. Several polymers, especially those with hydrophobic domains as their main form of holding the gel together, display thermoreverse properties.

HOW DOES IT WORK?

One classic misconception is that upon heating polymers become hydrophobic. This is not necessarily the case as polymers have hydrophobic and hydrophilic domains at all times but the change in temperature affects the relative impact of the organized/disorganized state of the water molecules (see Pelton R. Poly(N-isopropylacrylamide) (PNIPAM) is never hydrophobic. J Colloid Interface Sci 2010;348(2):673–674). For a good theoretical background on the driving principles of temperature response, a simple thermally responsive polymer, PEG, can be used as a general model to apply to other systems. By itself, PEG displays a cloud point, a higher-level temperature at which the polymer precipitates out of water, which varies by molecular weight but is reported to be around 100–120°C. PEG has a unique interaction with water in that when in dissolved state (or in any state other than freshly dehydrated) each PEG chain monomer is tightly bound to 2–3 water molecules and the water molecules participating in this binding are highly organized leading to a decreased entropy relative to the entropy of free water. The binding energy is an enthalpic term (ΔH_f—heat of fusion). As temperature increases the total energy of the system (described as Gibbs free energy of the system ($G = H - TS$, G—Gibbs free energy, H—enthalpy, T—temperature, S—entropy)) favors the entropic term rather than the enthalpic term and the water molecules prefer to be in unorganized form in free solution rather than bound to the PEG chain (see Harris JM. Poly (ethylene glycol) chemistry: biotechnical and biomedical applications. *Springer* 1992).

This process is shown in Fig. 5.3.

Note that a key component of this interaction is also the presence of hydrophobic groups. Although the hydrogen-bonding status between water and the polymer at increased temperature is affected negatively, the hydrophobic interactions are affected very little. When these are present the strength of the hydrophobic interactions becomes greater than the hydrogen bonding, thus causing the polymer chains to bind to each other.

THERMPOLYMERS FOR 3D PRINTING

Recent research indicates great potential for use of 3D printing and scaffolds for tissue repair and other uses. Bioprinting, which is the printing of living cells in a specific pattern, is a state-of-the-art method and has the potential to fabricate living organisms.

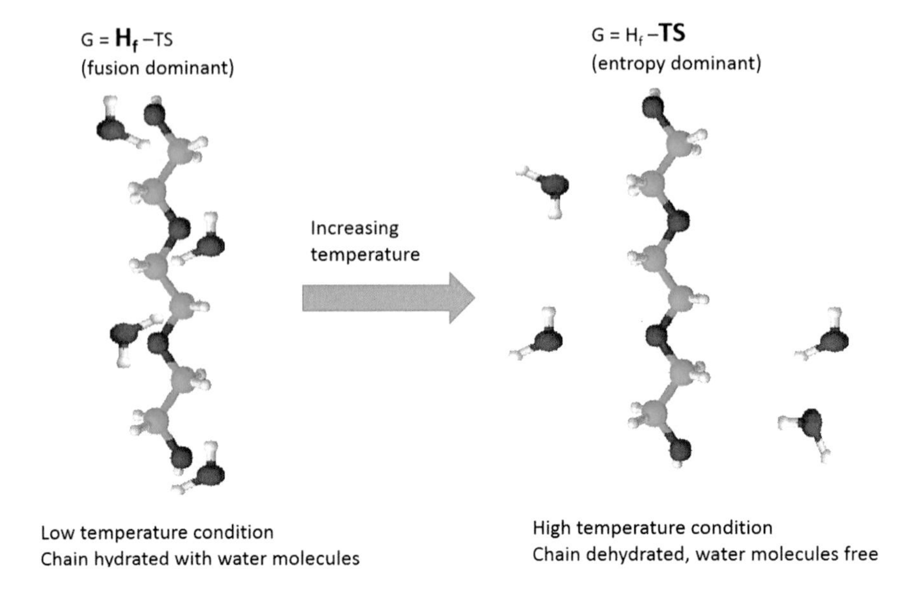

Figure 5.3 Schematic *of change in hydration status with heating.*

Bioprinting systems are classified into three different categories based on the methods used:

1. Laser-based systems processing 2D cell patterning. Laser direct write makes precise patterns of viable cells. These cells are suspended in solution in donor slides and are transported to a collector utilizing the laser energy. The laser pulse creates a bubble, which creates shockwaves. The shockwaves push the cells toward the collector on petri dishes.
2. Inkjet-based bioprinting uses living cells that are printed in the form of droplets through cartridges. This method enables printing either single cells or aggregate cells, depending on the process parameters.
3. Extrusion-based printing is another technique to print living cells. It is "the extrusion of continuous filaments made of biomaterials."

Each of these processes comes with its own advantages as well as limitations:

1. Laser-based systems have high resolution and allow precise patterning of living cells.
2. Inkjet-based systems are favored for cell encapsulation because of their versatility and affordability. The surfaces where the cells are printed and patterned do not have to be 2D. Drawbacks to this method include cell damage and death as well as cell sedimentation

and aggregation because of the small orifice diameter. The structural integrity of the printed structure is another concern.

3. The extrusion-based method yields much better structural integrity, and it is also the most convenient method to quickly make porous 3D structures. Limitations of this method also exist such as shear-stress-induced cell deformation and limited material selection due to the need for rapid cell encapsulation [68].

MEDICAL IMPLANTS

Patients have individual needs based upon their own anatomy and genetic makeup. Advanced manufacturing is very suitable for fabrication of personalized implants and devices. Patterning technologies can be designed to copy the surroundings and the regulatory microenvironment of cells in vivo and to modify the microenvironment in order to study the cellular response. Two-dimensional techniques have proved to be inadequate for some of the newer challenges of cell biology, biochemistry, and in pharmaceutical assays.

Three-dimensional structures are very important for in vitro experiments, which has been shown by multiple studies. For example, "hepatocytes retain many of their liver-specific functions for weeks in culture in-between two layers of collagen gel, whereas they lose many of these functions within a few days when cultured as a monolayer on the same gel." Advanced manufacturing techniques have been developed or modified to include cells in the fabrication process. These include biolaser-printing, stereolithography, and robotic dispensing (which is also known as 3D fiber plotting (3DF) or bioplotting). Rapid advancements in this field are proven by the establishment of the journal *Biofabrication* and the establishment of the International Society of Biofabrication in 2010.

Scaffolds

Additive tissue manufacturing is still in its infancy. Many biodegradable materials have been manufactured and employed to design and fabricate scaffolds and matrices including polymers (both natural and synthetic), ceramics, and composites. However, these materials usually require process parameters that are not conducive to direct inclusion on the cells. Therefore hydrogels are gaining the most interest in the manufacturing of tissues., since they are polymeric networks that absorb water while remaining insoluble and preserving their characteristic 3D structure. Hydrogels can do this due to the large number of physical or chemical links between the polymer chains. Hydrogel structures with viable cells have only been manufactured with simple and isotropic designs limited to only a few millimeters. The imposed requirement for mechanical properties are "self-supporting" and "handleable."

One of the factors that determine the biocompatibility of hydrogels is hydrophilicity, which makes hydrogels attractive for use in medicine and pharmacy as drug and cell carriers. One disadvantage of hydrogels is that their mechanical strength isn't as high as load-bearing tissues, which limits their use in advanced tissue manufacturing to larger scales. Obvious ways to increase the strength and modulus of the gels are to increase the polymer concentration and cross-link density.

Nanocomposite gels are a class of hydrogels that exhibit mechanical properties superior to conventional hydrogels. There have also been advancements in the strength of hydrogels outside biomedical engineering with novel chemical structures exhibiting improved mechanical properties, with increased toughness and strength while still containing high water-volume fractions. Advanced manufacturing techniques also offer more control over the scaffold architecture and range of materials that can be processed. Scaffolds for tissue engineering are usually prepared from ceramics, polymers, or a composite of the two [24–30].

BIOBOTS

Cells, tissues, and organs are not the only things on the horizon for bioengineers. Researchers have also created biobots, part gel, part muscle, which are expected to someday travel within the body to sense toxins and deliver medication. These biobots are made from heart muscle cells and work is being conducted to regulate the muscle contractions.

POLOXAMER 407

Rheology performed on AR550 (TA instruments) with a 60 mm 2 degree cone on 20% w/v polymer in PBS dissolved over 2 days with shaking at 4°C. Viscosity of solution at 0.1 (second^{-1}) and 5°C was measured (1 minute peak hold, 5 second test intervals). Rheology performed by oscillating at constant 6.283 rad/second, 0.1% strain, in increments of 2.5°C ranging from 5°C to 45°C with 3 minutes of temperature equilibration at each point (Figs. 5.4A–C).

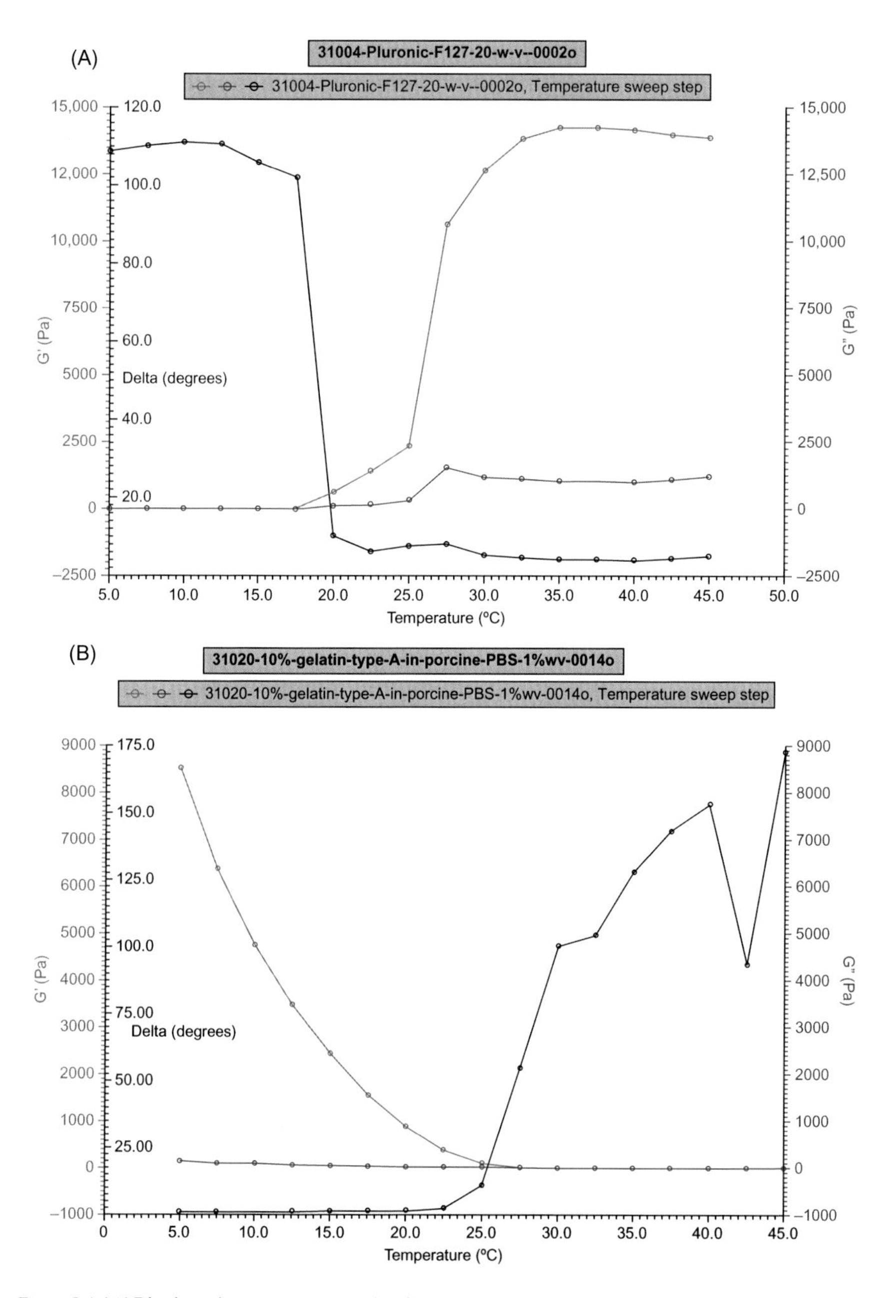

Figure 5.4 (A) Rheology chart, *viscosity 20% w/v solution at 5° C | 4.593 mPa·s; (B)* Rheology chart, *viscosity 10% w/v solution at 45° C | Below detectable limit; (C)* Rheology chart, *viscosity 10% w/v solution at 5° C | 19.85 Pa·s.*

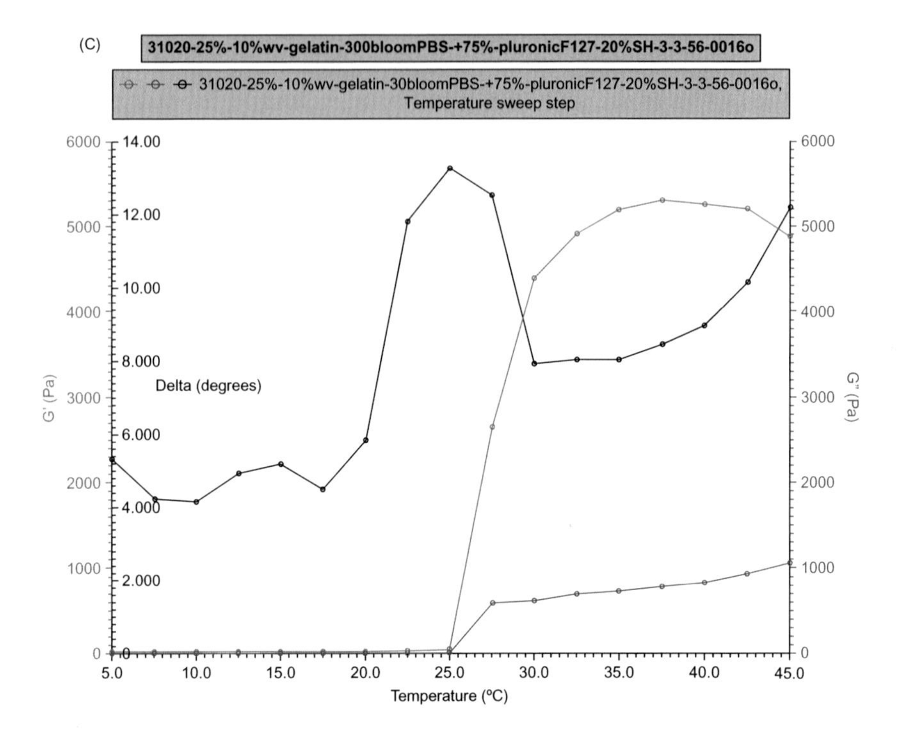

Figure 5.4 (Continued)

GELATIN TYPE A 300 BLOOM

Rheology performed on AR550 (TA instruments) with 60 mm 2 degree cone on 10% w/v polymer in PBS dissolved at 60°C. Viscosity of solution at 0.1 (second^{-1}) and 5°C was measured (1 minute peak hold 5 second test intervals). Rheology performed by oscillating at constant 6.283 rad/second, 0.1% strain, in increments of 2.5°C ranging from 45°C to 50°C with 3 minutes of temperature equilibration at each point.

Gelatin Type A 300 Bloom + Poloxamer 407

Rheology performed on AR550 (TA instruments) with 60 mm 2 degree cone. Solution prepared by mixing 3 mL 20% w/v poloxamer407 in PBS along with 1 mL of the 10% Gelatin 300 Bloom-type A solution. Viscosity of solution at 0.1 (sec^{-1}) and 5°C was measured (1 minute peak hold 5 second test intervals). Rheology performed by oscillating at constant 6.283 rad/second, 0.1% strain, in increments of 2.5°C ranging from 5°C to 45°C with 3 minutes of temperature equilibration at each point [31–43,68].

4D Printing

The next generation of 3D bioprinting is called 4D bioprinting. A definition of 4D bioprinting was presented by Dr. William Whitford in last month's "Another Perspective on 4D Bioprinting." In the article Dr. Whitford explains that 4D bioprinting is "the printing of smart, environmentally responsive biological structures, tissues and organs. 4D bioprinting begins with the printing of multiple cells or biological matrices resulting in structures that undergo subsequent designed and anticipated (not spontaneous) but self-originated development in response to an environment."

REFERENCES

[1] Wang Z, et al. A simple and high-resolution stereolithography-based 3D bioprinting system using visible light crosslinkable bioinks. Biofabrication 2015;7:045009.

[2] Jana S, Lerman A. Bioprinting a cardiac valve. Biotech Adv 2015;33(8):1503–21.

[3] Pereira RF, Bartolo PJ. 3D bioprinting of photocrosslinkable hydrogel constructs. J Appl Polym Sci 2015;132(48):132–4.

[4] Hribar CC, Soman P, Warner J, Chung P, Chen S. Light-assisted Direct-Write of 3D Functional Biomaterials. Lab Chip 2014;14:268–75.

[5] Huang TQ, et al. 3D printing of biomimetic microstructures for cancer cell migration. Biomed Microdevices Feb 2014;16(1):127–32.

[6] Nicodemus GD, Bryant SJ. Cell encapsulation in biodegradable hydrogels for tissue engineering applications. Eng Part B 2008;14(2):149–65.

[7] Soman P, et al. Cancer cell migration within 3D layer-by-layer microfabricated photocrosslinked PEG scaffolds with tunable stiffness. Biomaterials 2012;33:7064–70.

[8] Fairbanks BD, et al. Photoinitiated polymerization of PEG-diacrylate with lithium phenyl-2,4,6-trimethylbenzoylphosphinate: polymerization rate and cytocompatibility. Biomaterials Dec 2009;30(35):6702–7.

[9] Kesti M, et al. A versatile bioink for three-dimensional printing of cellular scaffolds based on thermally and photo-triggered tandem gelation. Acta Biomater 2015;11:162–72.

[10] Duan B, et al. 3D bioprinting of heterogeneous aortic valve conduits with alginate/gelatin hydrogels. J Biomed Mater Res A June 2013;101(5):1255–64.

[11] Khalil S, et al. Bioprinting endothelial cells with alginate for 3D tissue constructs. J Biomech Eng October 2009;131(11):111002.

[12] Kundu J, et al. An additive manufacturing-based PCL-alginate-chondrocyte bioprinted scaffold for cartilage tissue engineering. Tissue Eng Part B 2008;14(2):1286–97.

[13] Poldervaart MT, et al. Sustained release of BMP-2 in bioprinted alginate for osteogenicity in mice and rats. PLoS One August 2013;8(8):e72610.

[14] Song SJ, et al. Sodium alginate hydrogel-based bioprinting using a novel multinozzle bioprinting system. Artif Organs 2011;35(11):1132–6.

[15] Zhang Q, et al. Review on biomedical and bioengineering applications of cellulose sulfate. Carbohydr Polym November 2015;132:311–22.

[16] Mohite B, et al. A novel biomaterial: bacterial cellulose and its new era applications. Biotechnol Appl Biochem 2014;61(2):101–10.

[17] Kajsa M, et al. 3D bioprinting human chondrocytes with nanocellulose-alginate bioink for cartilage tissue engineering applications. Biomacromolecules 2015;16(5):1489–96.

[18] Adam R, et al. 3D bioprinting of carboxymethylated-periodate oxidized nanocellulose constructs for wound dressing. Biomed Res Inter 2015;2015:925757.

[19] Levato R, et al. Biofabrication of tissue constructs by 3D printing of cell-laden microcarriers. Biofabrication 2014;6:035020.

[20] Visser J, et al. Biofabrication of multi-material anatomically shaped tissue constructs. Biofabrication 2013;5:035007.

[21] Kirchmajer D, et al. An overview of the suitability of hydrogel-forming polymers for extrusion-based 3D-printing. J Mater Chem B 2015;3:4105–17.

[22] Lee Y-B. Bioprinting of collagen and VEGF-releasing Fibrin Gel Scaffolds for Neural Stem Cell Culture. Exp Neurol 2010;223:645–52.

[23] Ahmed TA, et al. Fibrin: a versatile scaffold for tissue engineering applications. Tissue Eng Part B 2008;14(2):199–215.

[24] Hinton TJ, et al. Three-dimensional printing of complex biological structures by freeform reversible embedding of suspended hydrogels. Sci Adv 23 Oct 2015;1(9):e1500758.

[25] Organ Procurement and Transplantation Network. U.S. Department of Health & Human Services. n.d. http://www.biega.com/3D-printing.shtml.

[26] Zhang X, Zhang Y. Tissue engineering applications of three-dimensional bioprinting. Cell Biochem Biophys 2015;72(3):777–82.

[27] New Findings on Tissue Engineering Described by Investigators at University of Iowa (a Hybrid Bioprinting Approach for Scale-Up Tissue Fabrication) report. Blood Weekly 2014;78. Print.

[28] Murphy SV, Atala A. 3D bioprinting of tissues and organs. Nat Biotechnol 2014;32(8):773–85.

[29] Miller JS, Stevens KR, Yang MT, Baker BM, Nguyen D-HT, Cohen DM, et al. Rapid casting of patterned vascular networks for perfusable engineered three-dimensional tissues. Nat Mater 2012;11(9):768–74.

[30] Lang-8. Bioprinting: Brief Timeline and Outlook.

[31] http://lang8.com/1017887/journals/107384577998436484655804992513119667379.

[32] Kolesky DB, Truby RL, Gladman AS, Busbee TA, Homan KA, Lewis JA. 3D Bioprinting of vascularized, heterogeneous cell-laden tissue constructs. Adv Materials 2014;26 (19):3124–30.

[33] Harris W. How 3-D bioprinting works. HowStuffWorks, n.d. Web.

[34] Gu Q, Hao J, Lu Y, Wang L, Wallace GG, Zhou Q. Three-dimensional bio-printing. Sci China: Life Sci 2015;58:411–19.

[35] http://geeksandknots.blogspot.com/2013/04/print-what-servin-up-some-frankenstein.html.

[36] Fermeiro JBL, Calado MRA, Correia IJS. State of the art and challenges in bioprinting technologies, contribution of the 3D bioprinting in Tissue Engineering. Bioengineering 2015;1–6. Available from: http://dx.doi.org/10.1109/ENBENG.2015.7088883.

[37] Paulsen SJ, Miller JS. Tissue vascularization through 3D printing: will technology bring us flow? Dev Dynam 2015;244:629–40.

[38] Hoch E, Tovar GEM, Borchers K. Bioprinting of artificial blood vessels: current approaches towards a demanding goal. Eur J Cardio-Thoracic Surg 2014;46(5):767–78.

[39] Bakhshinejad A, D'Souza RM. A brief comparison between available bio-printing methods. Great Lakes Biomedical Conference; 2015.

[40] Gao Q, He Y, Fu J, Ma L. Coaxial nozzle-assisted 3D bioprinting with built-in microchannels for nutrients delivery. Biomaterials 2015;61:203–15.

[41] Stieglitz LH, Gerber N, Schmid T, Mordasini P, Fichtner J, Fung C, et al. Intraoperative fabrication of patient-specific moulded implants for skull reconstruction: single-centre experience of 28 cases. Acta Neurochir (Wien) 2014;156(4):793–803.

[42] Kucukgul C, Ozler B, Inci I, Karakas E, Irmak S, Gozuacik D, et al. 3D bioprinting of biomimetic aortic vascular constructs with self-supporting cells. Biotechnol Bioeng 2014;112 (4):811–21.

[43] Findings on Tissue Engineering Detailed by Investigators at University of Iowa (bioprinting Technology: a Current State-of-the-Art Review). Blood Weekly 2014;54. Print.

[44] Hull CW. Apparatus for production of three-dimensional objects by stereolithography, 1986, March 11. Retrieved from http://www.google.com/patents/US4929402.

[45] Kang H-W, et al. A 3D bioprinting system to produce human-tissue constructs with structural integrity. Nat Biotechnol March 2016;34(3):313–22.

[46] Kolesky DB., et al., Three-dimensional bioprinting of thick vascularized tissues. PNAS Early Edition, 2016.

[47] Hong S, et al. Cellular behavior in micropatterned hydrogels by bioprinting system depended on the cell types and cellular interaction. J Biosci Bioeng August 2013;116 (2):224–30.

[48] Horvarth L, et al. Engineering an in vitro air-blood barrier by 3D bioprinting. Nat Sci Rep 2015;5:7974.

[49] Astashkina DWG, et al. Critical analysis of 3-D organoid in vitro cell culture models for high-throughput drug candidate toxicity assessments. Adv Drug Deliv Rev 2014;69:1–18.

[50] Nocera D, et al., Printing collagen 3D structures. In: VI Latin American Congress on Biomedical Engineering CLAIB 2014 (Vol 49), Parana, Argentina, IFMBE Proceedings, 2014, pp. 136–139.

[51] Park JY, et al. A comparative study on collagen type I and hyaluronic acid dependent cell behavior for osteochondral tissue bioprinting. Biofabrication 2014;6:035004.

[52] Lee V, et al. Design and fabrication of human skin by three-dimensional bioprinting. Tissue Eng Part C 2014;20(6):473–84.

[53] Nichol JW, et al. Cell-laden microengineered gelatin methacrylate hydrogels. Biomaterials 2010;31:5536–44.

[54] Schacht K, et al. Biofabrication of cell-loaded 3D spider silk constructs. Angew Cehm Int Ed 2015;54:1–6.

[55] Highley CB, et al. Direct 3D printing of shear-thinning hydrogels into self-healing hydrogels. Adv Mater 2015;27:5075–9.

[56] Pescosolido L, et al. Hyaluronic acid and dextran-based semi-IPN hydorgels as biomaterials for bioprinting. Biomacromolecules 2011;12(5):1831–8.

[57] Rodell CB, Mealy J, Burdick JA. Supramolecular guest-host interactions for the preparation of biomedical materials. Bioconjugate Chem 2015;26:2279–789.

[58] Burdick JA, et al. Controlled degradation and mechanical behavior of photopolymerized hyaluronic acid networks. Biomacromolecules 2005;6(1):386–91.

[59] Levesque S, et al. Macroporous interconnected dextran scaffolds of controlled pororsity for tissue-engineering applications. Biomaterials 2005;26:7436–46.

[60] Aubin H, et al. Directed 3D cell alignment and elongation in microengineered hydrogels. Biomaterials September 2010;31(27):6941–51.

[61] Nikkhah M, et al. Directed endothelial cell morphogenesis in micropatterned gelatin methacrylate hydrogels. Biomaterials December 2012;33(35):9009–18.

[62] Bertassoni L, et al. Direct-write bioprinting of cell-laden methacrylated gelatin hydrogels. Biofabrication 2014;6:024105.

[63] Bertassoni LE, et al. Hydrogel bioprinted microchannel networks for vascularization of tissue engineering constructs. Lab Chip 2014;14:2202–11.

[64] Malda J, et al. 25th anniversary article: engineering hydrogels for biofabrication. Adv Mater 2013;25:5011–28.

[65] Carrow JK, et al. Polymers for bioprinting. Essentials of 3D biofabrication and translation. Elsevier Inc; January 2015. p. 229–48.

[66] Boere K. Hybrid dual cross-linked hydrogels: injectable and 3D-printable biomaterials. Utrecht, the Netherlands: Utrecht Institute for Pharmaceutical Sciences; 2015.

[67] Seliktar D. Designing cell-compatible hydrogels for biomedical applications. Science 2012;336:1124–8.
Shikanov A, et al. Characterization of the crosslinking kinetics of multi-arm poly(ethylene glycoL) hydrogel formed via Michael-type addition. Soft Matter 2016;12:2076–85.

[68] Ozbolat IT, Yu Y. Bioprinting toward organ fabrication: challenges and future trends. IEEE Trans Biomed Eng 2013;60(3):691–9.

CHAPTER 6

CT Scans Function Like a CAD Design

Clinical professionals utilize current tracers and radiological techniques; instead of trying to create an organ or tissue model from the ground up, researchers and engineers can use a CT scan or MRI to create a 3D model to print. For example, when creating a 3D printed model of a heart for use in surgery, researchers can use a CT scan to design the model.

Digital imaging and communication in medicine (DICOM) [12,13] is a standard used in medical imaging and intended for storing, viewing, printing, and sharing graphics, as well as provides information about patients, examination processes, hospitals, medical equipment manufacturers, etc. [1–6]. Based on open systems interconnection (OSI) (an international organization for standardization (ISO) standard), DICOM is associated with major medical equipment brands and software backed by this equipment. There are two information levels defined by the DICOM standard: the object-oriented tag-structured DICOM file and the DICOM network protocols [7–11]. Software supporting the DICOM format includes 3DSlicer, Ginkgo CADx, OsiriX, CDCM, MicroDicom, XnView, and Orthanc (Fig. 6.1).

This chapter discusses how embedded data from CT or MRI scans is used to create a 3D printable model. Different parts of the anatomy will be used as an example, with embedded data in the DICOM format (Fig. 6.2).

Download and install the free or pro version of the software from http://www.osirix-viewer.com/Downloads.html. The extra files and examples are helpful to practice with prior to use with clinical DICOM images (Fig. 6.3).

GET THE SCAN DATA

Clinical body scanners produce image files in the DICOM format, which can usually be found in a folder filled with a bunch of files with the .dcm extension (Fig. 6.4).

Bioprinting. DOI: http://dx.doi.org/10.1016/B978-0-12-805369-0.00006-7

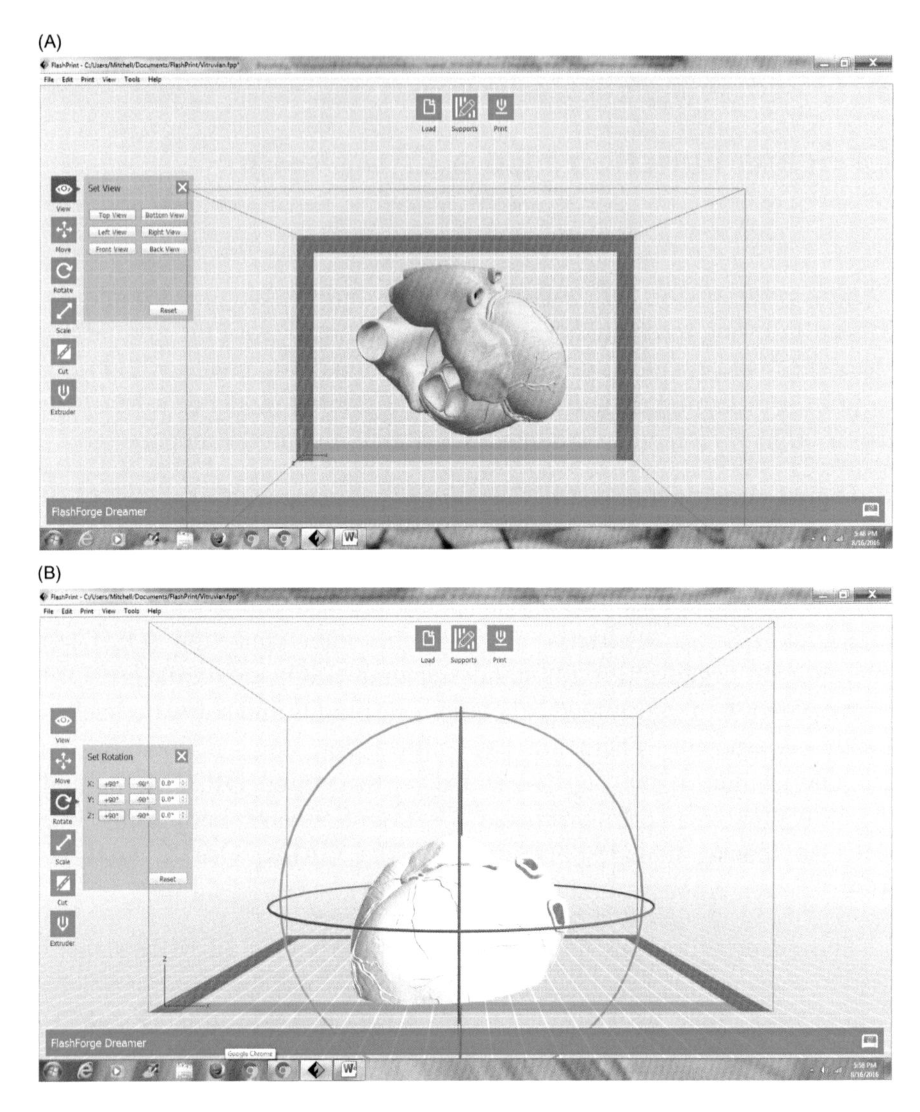

Figure 6.1 (A) CT scan of a heart cut in half using FlashPrint version 3.1.2.0 software (top view); (B) CT scan of heart cut in half using FlashPrint (front view).

GET THE SCAN DATA INTO OSIRIX

Launch Osirix. In the top-left corner of the screen click the Import button and browse to the folder that has your DICOM data.

Tip: When the FOLDER name is selected in the dialog, press the Open button and Osirix will load all the files contained in the folder.

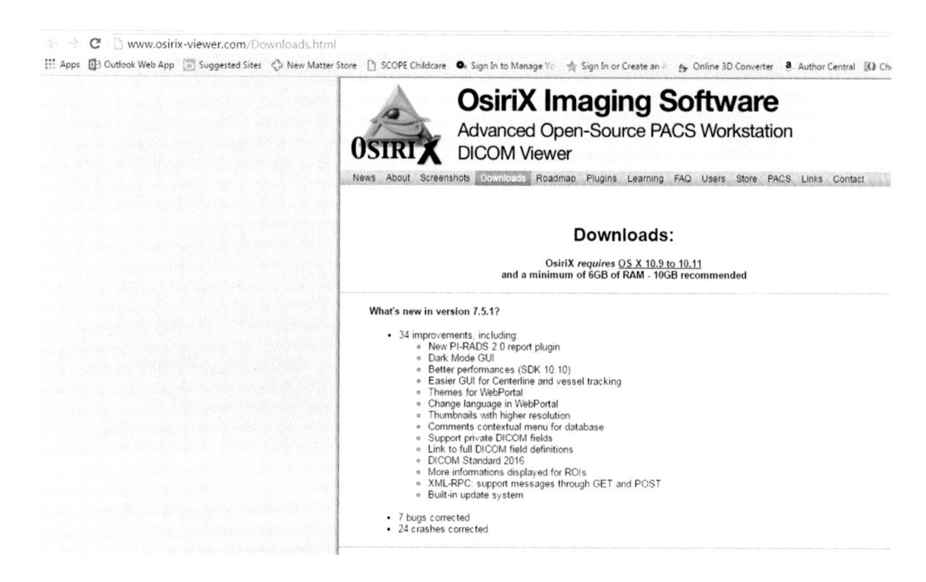

Figure 6.2 System requirements for Osirix imaging software:

- *OS X 10.9 to OS X 10.11*
- *Intel processor*
- *For best performance:*
- *6 GB of RAM if you plan to open more than 800 images (CT and MRI, PET-CT)*
- *8 GB of RAM for more than 1500 images (multislice CT and PET-CT) with OsiriX MD*
- *12 GB of RAM for more than 3000 images (cardiac or functional imaging) with OsiriX MD*

Another tip: When it asks you whether to Copy or Link the files, choose Copy, so Osirix will copy the files to its own database. You can then delete the downloaded .zip file and the copies of the DICOM data (Fig. 6.5).

CLINICAL IMAGE DATA IN 2D/3D VIEW

Once the files have been imported, select your data set in the top panel and then double-click the thumbnail image. It will open in 2D view, but you can scroll to view all of the various "slices" for the DICOM image.

To open it in 3D view, on the top toolbar look for the button with a gear labeled 2D/3D. In the drop-down menu, choose 3D Surface Rendering (Fig. 6.6).

OsiriX Software

OsiriX MD 64-bit, FDA-Cleared / CE certified for medical imaging, full version with all features, load unlimited number of images:

OsiriX MD

here

Download **OsiriX Lite (free lite version)**: OsiriX Lite.dmg

here

Download the OsiriX application for the **iPhone**, **iPod Touch** and **iPad**: OsiriX Mobile (Online Manual)

here

Extra-Files and Examples

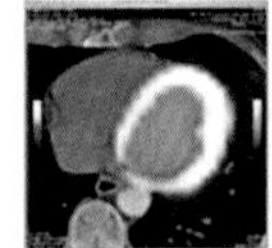
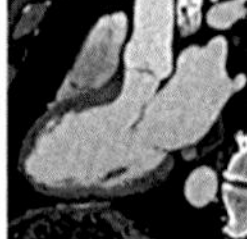
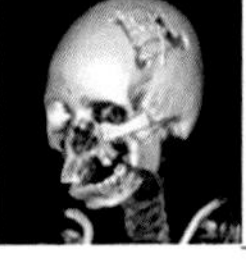
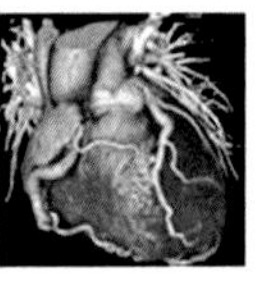

DICOM sample image datasets Web Site

Figure 6.3 Osirix download page.

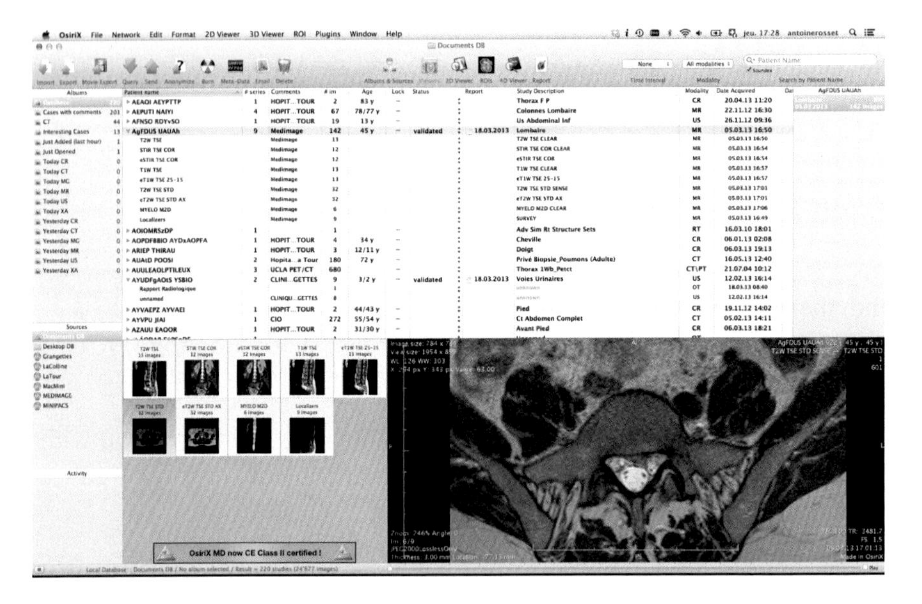

Figure 6.4 Osirix software on Mac OS computer.

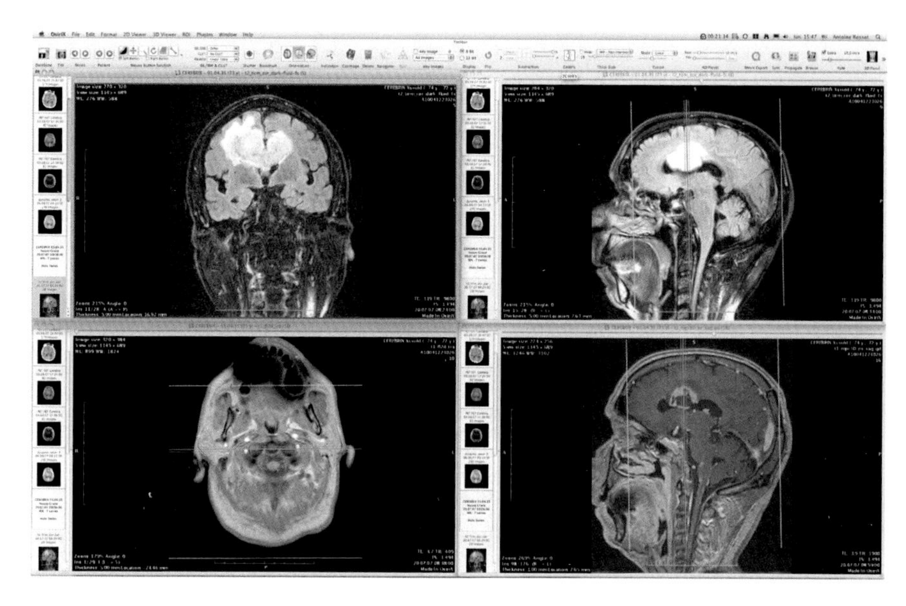

Figure 6.5 2D view of CT scan DICOM image in Osirix software.

EXPORT OF DICOM IMAGE SCAN AS 3D PRINTABLE MODEL

In the top toolbar, there is a button with a gear labeled "Export 3D-SR." Clicking this button will open a menu with different export format options, which will allow you to select either OBJ or STL formats to export for 3D printing/bioprinting.

3D Printing/Bioprinting the Scanned Image

Load the model as shown in Figs. 6.1A and B into FlashPrint (or whatever print utility/slicer you use), and set to print. The resulting model is shown in Fig. 6.7.

OTHER USES FOR DICOM IMAGES

DICOM images can also be converted to standard JPEG images, a standard image file format used by digital cameras and found widely on the Internet. When you use the JPEG format, you lose some image quality because of compression, but you also have a file that is much easier to archive, send, and publish.

MicroDicom is a free PC utility for converting DICOM images (extension *.dcm) to JPEG images and vice versa (Fig. 6.8).

(A)

(B)

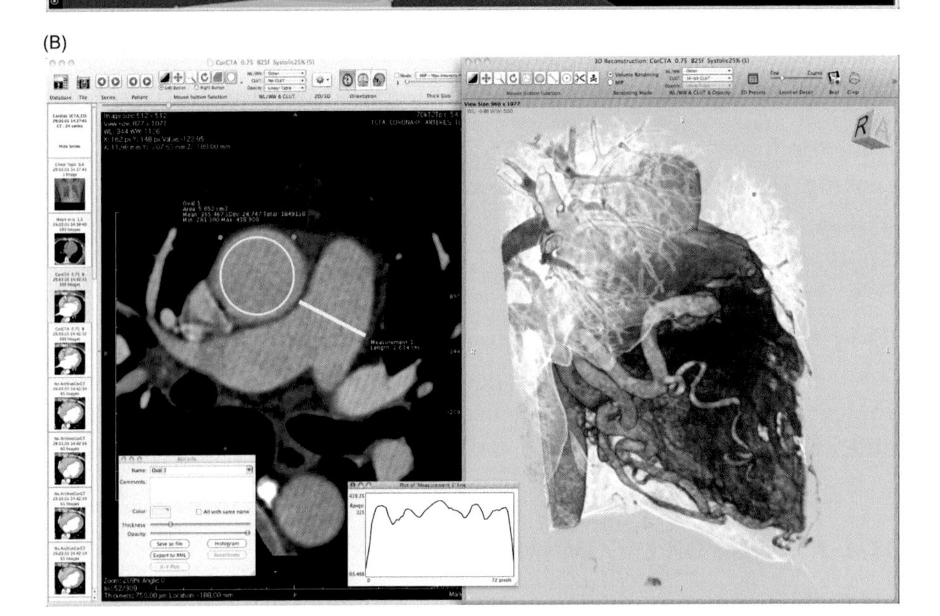

Figure 6.6 (A) 3D reconstruction of CT scan DICOM images (brain); (B) 3D reconstruction of CT scan DICOM images (heart); (C) 3D reconstruction of CT scan DICOM images (pelvis, spinal column, kidneys, and lower rib cage); (D) 3D reconstruction of CT scan DICOM images (skull and brain).

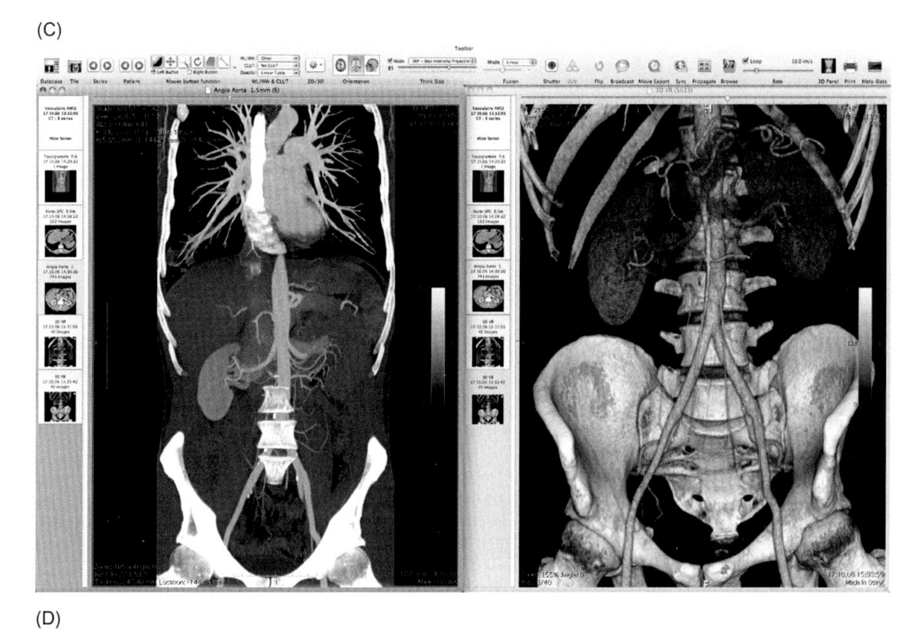

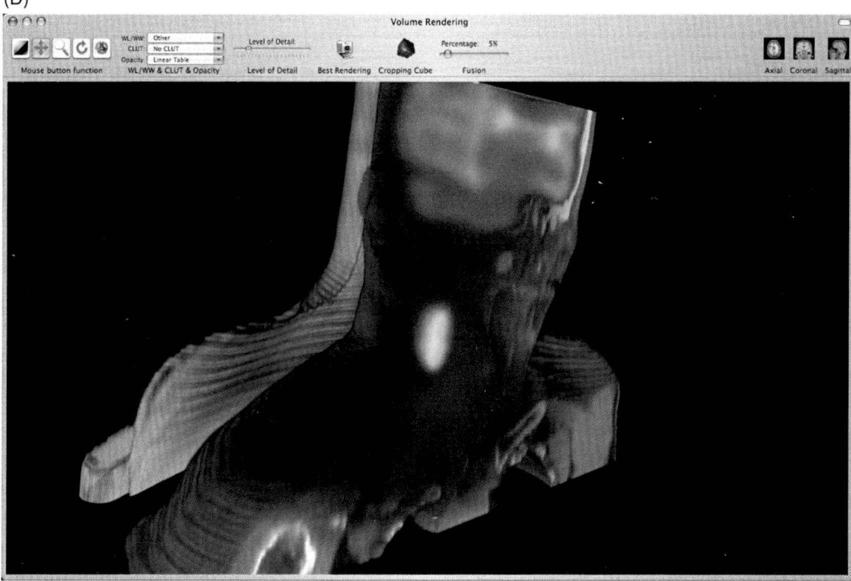

Figure 6.6 (Continued)

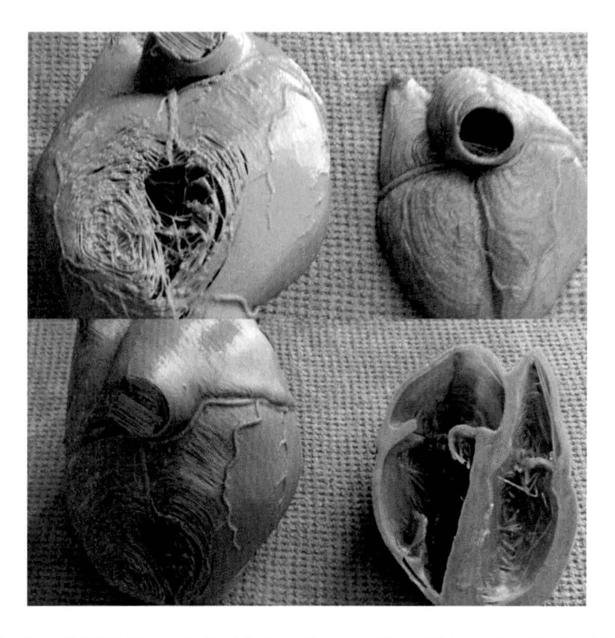

Figure 6.7 Results from DICOM image in FlashPrint software in Figs. 6.1A and B. 3D printed heart with bronze PLA filament material on the Flashforge Dreamer from the CT scan DICOM image.

MICRODICOM SHELL EXTENSION

This is a Windows shell extension that makes DICOM files easy to view in Windows Explorer.

Features:

View DICOM images in Windows Explorer
View DICOM tags in Windows Explorer on mouse over
Copy DICOM image to clipboard
Supports Windows 8 and Windows 8.1
Universal installer for x86 and x64 platforms

Most available commercial software for manipulating DICOM images is expensive to buy alone or distributed with expensive medical machinery. On the other hand, free software packages are often nonintuitive and with limited functionality. This means users often need more than one package to achieve feasible results.

Figure 6.8 MicoDicom DICOM viewer software for PC
System requirements:
Windows 98/Me/NT4/2000/XP/Vista
Processor 486DX or higher
32 MB RAM or more
11 MB free disk space
Recommended:
Windows XP
Processor Pentium II
128 MB RAM
Internet Explorer V5 or later

That's where MicoDicom DICOM viewer software comes in—it is equipped with most common tools for manipulation of DICOM images and it has an intuitive user interface. It is also free to use and accessible to everyone.

MicroDicom conversion is also very intuitive. All you have to do is open the software and browse for the selected DICOM images on the server or local folder (Fig. 6.9).

Next, you click the Export to Image Icon in the Main menu in the upper toolbar (Fig. 6.10A). As shown in Fig. 6.10B, the pop-up window allows you to customize how the file is exported (image format, size, overlay, etc.)

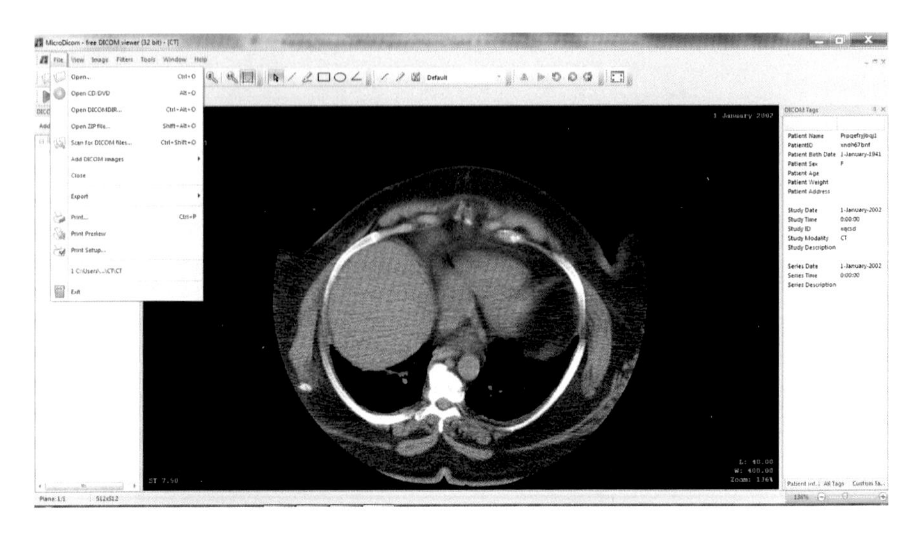

Figure 6.9 Viewing DICOM files using the MicroDicom software viewer.

PREPARING FOR 3D PRINTING

The JPEG image created from the original CT DICOM image is now analyzed in FlashPrint, a 3D slicer and viewer program (Fig. 6.11).

Once again there is a pop-up window to customize viewing (Fig. 6.12).

The customizations created a planar topographical relief of the image which can show any e vertical and horizontal dimension of the lesions or pathological incidence from the now 3D image.

Fig. 6.13 shows some pathology in the groin/pelvic area, rib cage, and scapula.

Figs. 6.14A and B show the results after analysis by FlashPrint.

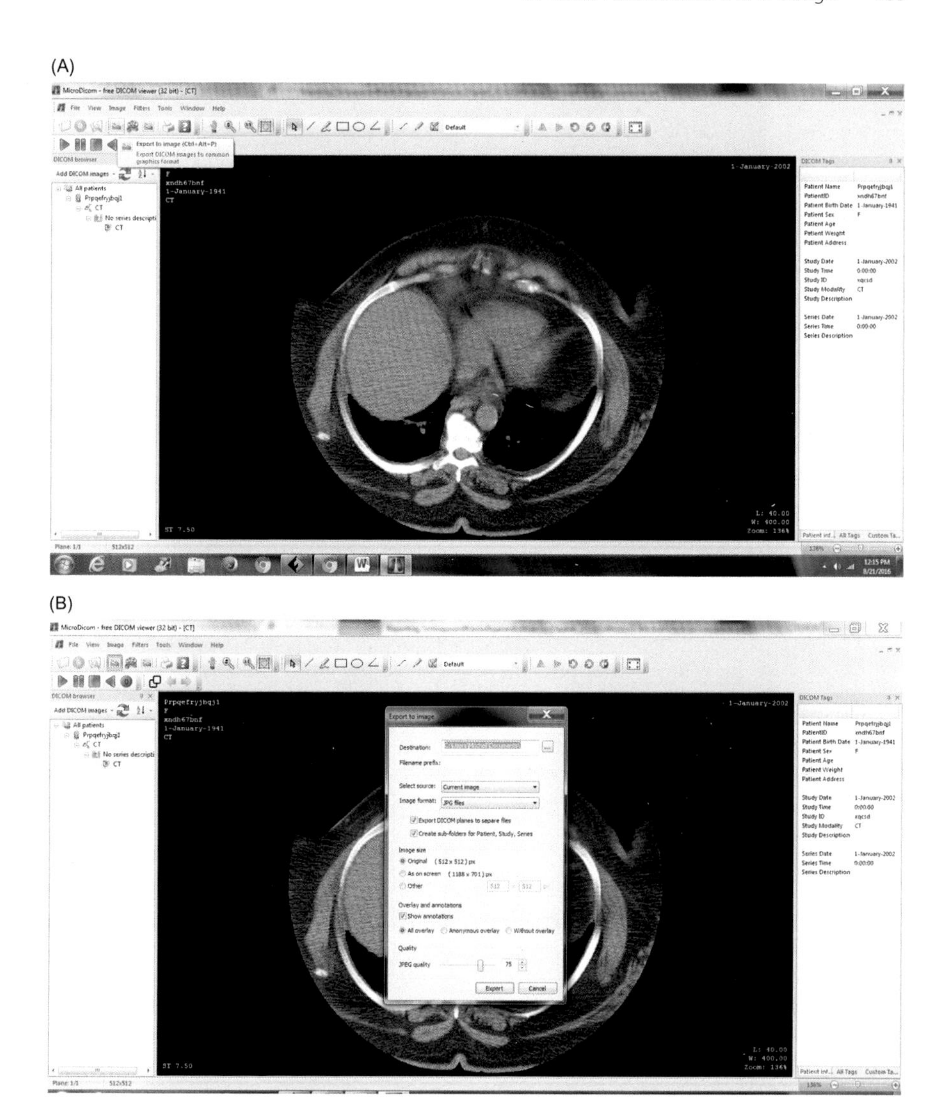

Figure 6.10 (A) Export CT DICOM image from the main menu bar; (B) Export of CT DICOM image customization pop-up window. The default values are used here. For better quality the pixels were increased, and the JPEG quality was increased from 75% to 100%; (C) Updated customizations for CT DICOM image export. Exported image (Fig. 6.10D); (D) Exported CT DICOM scan converted to high-quality JPEG image for 3D printing.

(C)

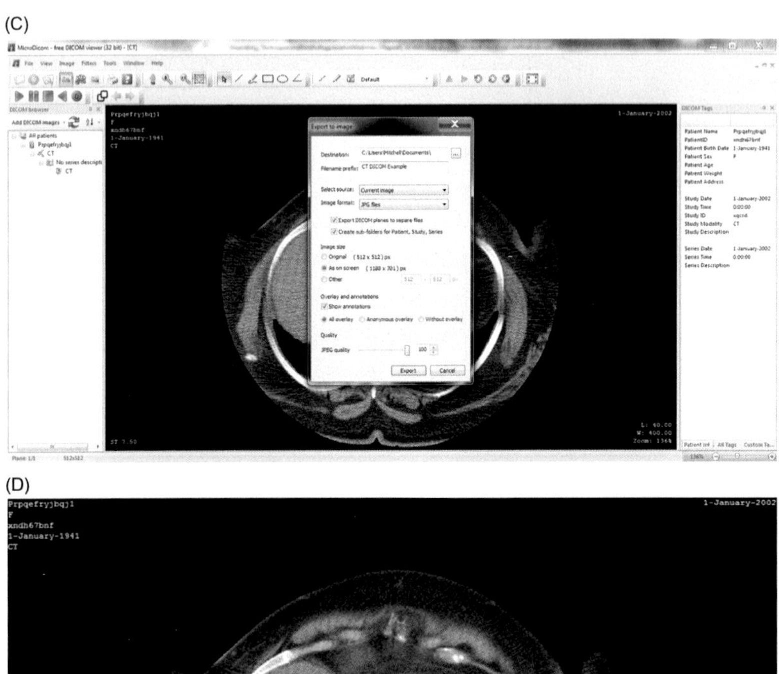

(D)

Figure 6.10 (Continued)

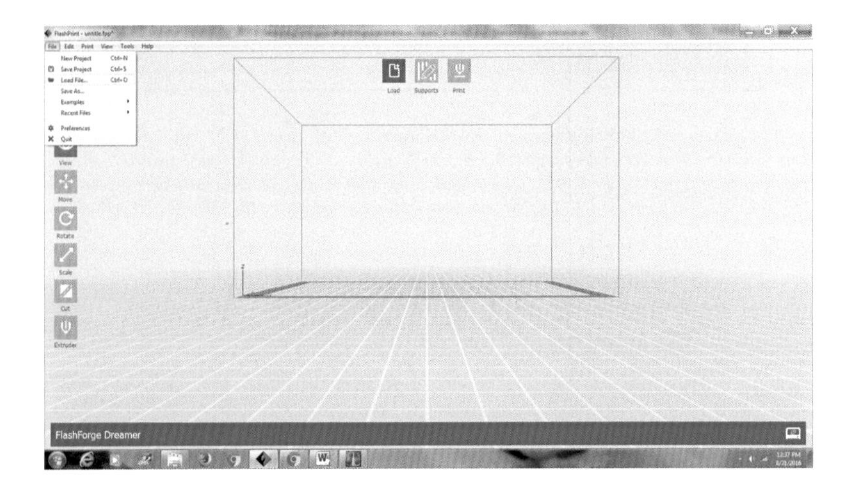

Figure 6.11 Loading the CT DICOM file into FlashPrint.

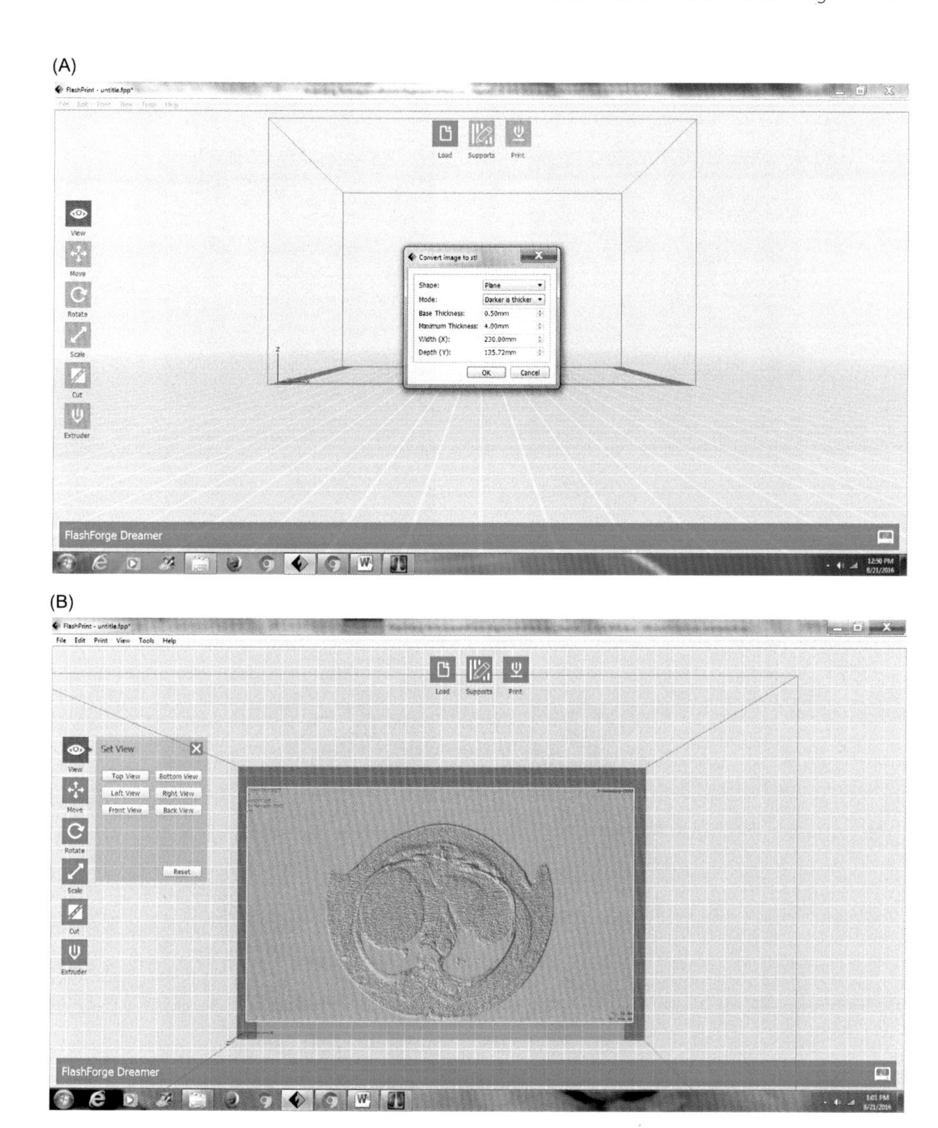

Figure 6.12 (A) FlashPrint customization of CT DICOM scan image to JPEG image; (B) FlashPrint customization of CT DICOM scan image to JPEG image result.

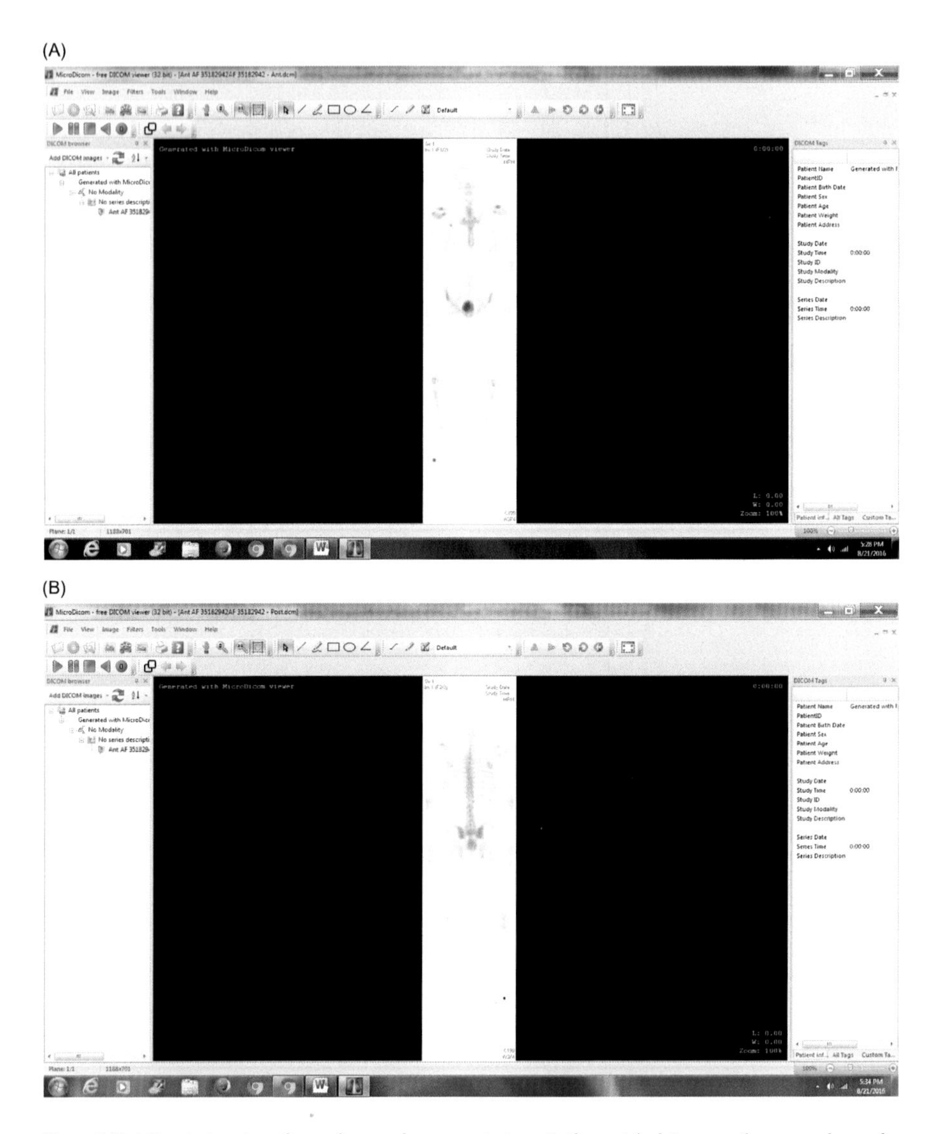

Figure 6.13 (A) anterior view of some bone and organ metastases in the groin/pelvic area, rib cage, and scapula; (B) posterior view (above) showing some bone and organ metastases in the groin/pelvic area, rib cage, and scapula and the resulting converted JPEG image.

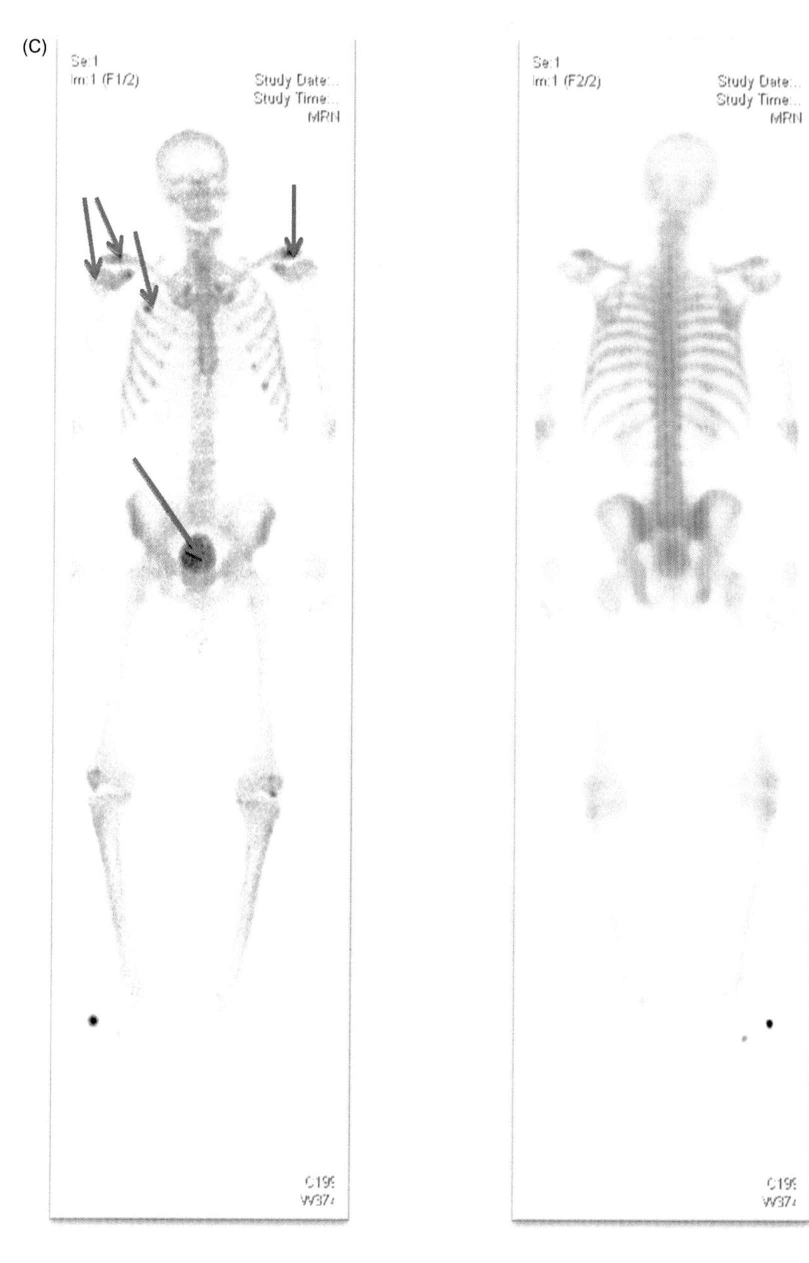

Figure 6.13 (Continued)

(A)

(B)

Figure 6.14 (A) Anterior view of male patient with metastases located in bone and organs; (B) 3D print result of the anterior view of male patient with metastases located in bone and organs using Saintsmart silver PLA filament and Flashforge Dreamer 3D printer.

REFERENCES

[1] DICOM brochure, nema.org.

[2] MEMBERS of the DICOM STANDARDS COMMITTEE.

[3] http://www.nema.org/About/Pages/Members.aspx.

[4] Kahn Jr CE, Carrino JA, Flynn MJ, Peck DJ, Horii SC. DICOM and radiology: past, present, and future. J Am Coll Radiol 2007;4:652–7. Available from: http://dx.doi.org/10.1016/j.jacr.2007.06.004.

[5] http://www.astm.org. If a Picture Is Worth 1,000 Words, then Pervasive, Ubiquitous Imaging Is Priceless.

[6] IHE Profiles.

[7] http://www.nema.org. Industrial Imaging and Communications Section.

[8] Shiroma JT. An introduction to DICOM. Vet Med 2006;19–20 Retrieved from: http://0-search.proquest.com.alpha2.latrobe.edu.au/docview/195482647?accountid=12001.

[9] DICOM Strategy Document.

[10] Mustra M, Delac K, Grgic M. Overview of the DICOM Standard (PDF). ELMAR, 2008. 50th International Symposium. Zadar, Croatia; September 2008, pp. 39–44. ISBN 978-1-4244-3364-3.

[11] Flanders AE, Carrino JA. Understanding DICOM and IHE. Semin Roentgenol 2003;38:270–81.

[12] http://medical.nema.org/Dicom/2011/11_14pu.pdf.

[13] Clunie D, Cordonnier K. Digital Imaging and Communications in Medicine (DICOM) – Application/dicom MIME Sub-type Registration. IETF. RFC 3240; February 2002. Retrieved 2014-03-02.

CHAPTER 7

Additive Manufacturing and 3D Bioprinting for Pharmaceutical Testing

The use of polymers as biomaterials has been the subject of intense investigation over the past 50 years [1,2]. Different chemical structures and functional groups in such polymers govern their morphology and properties and allow precise control of the creation of desired molecular architectures for a wide range of applications in the biomedical field. For example, biocompatible polymers have been used successfully as artificial organs and drug delivery systems [3,4]. However, it is to be noted that the degree of success in such applications depends on the self-organization and biocompatibility of the formulated molecular architecture.

The biomaterials derived from polymers generally fall into two categories: naturally occurring and human-made synthetic materials. Collagens, alginate, and chitosan-based materials are the best examples of biomaterials derived from natural resources. The polymers derived from synthetic origins are divided into two classes: nonbiodegradable and biodegradable synthetic polymers. Recently, the biodegradable polymers have become highly important in the field of biomaterials and tissue engineering due to the avoidable additional surgery to remove the implants or scaffolds. Thus much attention needs to be given to the synthesis of biodegradable polymers.

In medical applications there is ongoing research and development (R&D) for the improvement of methodologies and devices for more efficient and effective processing of biomaterials. The outcome of such R&D has recently successfully been applied to treat many diseases [5–7]. Among the wide range of biomaterials that have been synthesized for potential use in medicine, the majority of these do not have suitable properties to interact effectively with biological tissues or cells. However, it is possible to improve their intrinsic properties using appropriate process engineering for optimum results. Crosslinking of

Biopringting. DOI: http://dx.doi.org/10.1016/B978-0-12-805369-0.00007-9

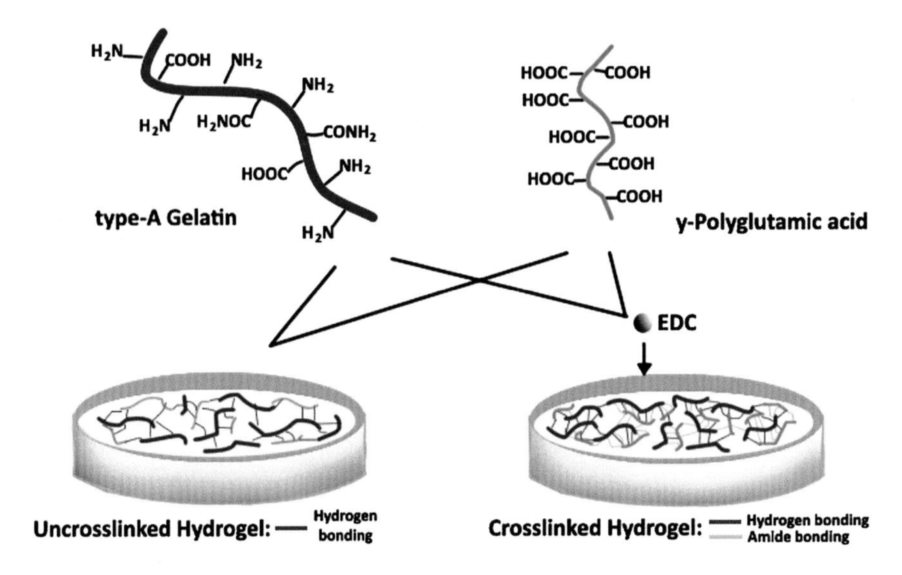

Figure 7.1 Schematic representations of the preparation process of two types of polymer hydrogels (HGs). HGs that are produced by covalent links between polymeric chains can be created by the use of reactive crosslinker(s) with or without initiators ("chemical" gels).

biopolymers is one example of process engineering that has provided a means to improve the quality of biomaterials for wider medical applications. For example, the crosslinked form of soft polymers, classified as hydrogels (HGs) [8], is a new generation of biomaterials that has demonstrated the ability to form scaffolds for a variety of uses, such as tissue engineering, delivery of active molecules, and biosensors and actuators. HGs are 3D structured polymeric materials, called "swell gels," that are formed via crosslinking reactions of polymers (Fig. 7.1).

The HGs can be synthesized with the required properties depending on the chemical structure, composition, and confirmation of starting materials, density of linking of polymer chains, hydrophobicity, and hydrophilicity for a particular biomedical application.

The 3D structural integrity and properties of HGs are mainly dependent on their method of preparation such as physical or chemical crosslinking reaction [3,4]. HGs from chemical crosslinking form permanent junction-type networks. The physical crosslinking of HGs allows forming transient junction-type networks, such as polymer chain entanglements, and physical HGs can be synthesized both from natural and synthetic polymers. Example HGs from natural polymers include collagen, gelatin, hyaluronic acid, chondroitin sulfate, chitin and chitosan, alginate, starch, cellulose, and their derivatives. HGs from natural polymers have many advantages over synthetically derived ones such as low toxicity and good biocompatibility because of their chemical structures and are very akin to the structure of glycosaminoglycan molecules present in the native extracellular matrix. HGs from synthetic polymers are prepared by chemical polymerization methods. Various types of monomers (e.g., acrylates, methacrylates, acrylamides, esters, carboxylic acid, and polyfunctional monomers) can be utilized for the preparation of synthetic HGs [9]. A detailed description of the preparation of HGs is beyond the scope of this review. This topic has been covered in-depth by several researchers [9–11].

In this review, we describe the recent developments of polymeric biomaterials and 3D structure generation by utilizing a variety of advanced techniques and methods with an emphasis on various types of tissue engineering. Several strategies for the 3D scaffold fabrication, which includes lithography and printing techniques, patterning by self-organization of polymers, self-assembling of peptides, and cellular compatibility of polymer-based biomaterials and HGs, are presented. The advantages and drawbacks of the 3D scaffold fabrication methods are also discussed. Additionally, we describe the applications of polymeric biomaterials and scaffolds in tissue engineering, particularly to cartilage, bone, and neural tissue regeneration. Furthermore, approaches to the incorporation of bioactive factor molecules in biomaterials via physical encapsulation and chemical crosslinking, their functions and specific applications in tissue regeneration, are also discussed.

TISSUE ENGINEERING

The objectives of tissue engineering are to replace, repair, or regenerate damaged tissues, or to create artificial tissues for transplantation, when normal physiologic reaction fails to take place and surgery becomes essential. Currently two different standards are used (see Figs. 7.2 and 7.3): autografts and allografts.

However, each has several limitations, including donor-site morbidity in the case of using autografts and the associated potential risk of disease transmission in the case of using allografts. Considerable research has

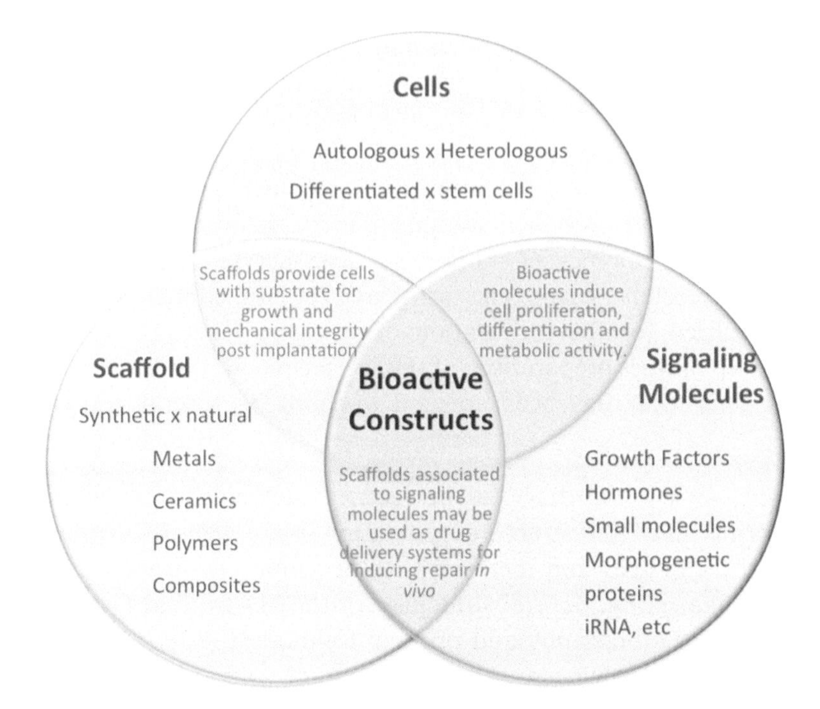

Figure 7.2 A schematic for developing tissue engineering strategies. In order to obtain tissue regeneration, cells, scaffolds, and signaling molecules may be introduced into the body alone or in association. Currently, the association of all three elements, composing bioactive constructs, is proposed to be the best option for tissue engineering.

	AUTOGRAFT	ALLOGRAFT	
PROS	No immunosuppression	2-mo immunosuppression	CONS
CONS	Cell culture takes 1-mo Graft delay → scar Elderly cells less regenerative High QA/QC costs for each individual graft Host limitation Infected/mutated host tissues may contaminate system	Good cells readily available Grafting within 12 hr ↛ scar Young cells rejuvenate Low QA/QC costs for 50 grafts at a time Unlimited supply Patients with infections and genetic diseases can be treated	PROS

Figure 7.3 Pros and cons of using and autograft versus an allograft.

been done worldwide to overcome the inherent limitations of current standards and to improve the biomedical technology by employing 3D biomaterial scaffold-based tissue engineering strategies. In the scaffold-based tissue engineering approach, it is essential that the interactions of 3D scaffold materials and cells take place by means of biocompatibility, cell adhesion, proliferation, growth, differentiation, and matrix deposition. Scaffolds must be designed with an appropriate surface chemistry and morphology to promote cellular functions and with sufficient structural and physical properties such as mechanical strength, porosity, and pore size. Such scaffolds can be fabricated from the original biodegradable and nonbiodegradable polymers. In the case of a biodegradable 3D scaffold, it must be designed in such a way so that it maintains structural integrity and functions and degrades in a controlled manner, until the new tissues are formed and the function continues.

While the technology is not yet advanced enough to create a full organ, the tissue samples are perfect for testing drugs and other medical advancements. Instead of having to use human beings or animals as guinea pigs for pharmaceutical testing, bioprinting can provide a much more cost-effective and ethical option, while still being accurate because the tissue samples are made from human cells.

There are several strategies in tissue engineering currently under investigation; some examples are schematically described in Fig. 7.4. Most of these utilize cells that are seeded onto 3D scaffolds. Scaffolds are generally designed to be fabricated with a wide range of properties to allow for cell–cell communication and migration, cell proliferation and differentiation, and to maintain the biocompatibility and structural integrity throughout the tissue regeneration process.

The methods of fabrication of biocompatible 3D scaffolds with appropriate architectures are divided into two areas: conventional and rapid prototyping. Conventional fabrication methods often do not provide sufficient physical and mechanical properties, and consequently these types of scaffolds undergo deformation because of cell motility. Rapid prototyping methods do not have such disadvantages and can provide all essential characteristics for specific tissue engineering applications. In fact, 3D nano/micro pattern scaffolds fabricated by rapid prototyping have shown a significant influence on cellular morphology, cell proliferation, and differentiation and also on the functioning of various cell types [12–14]. Scaffold fabrication by conventional methods includes phase separation [15], porogen leaching [16], gas foaming [17], fiber meshing [18], and supercritical fluid processing [19]. The second category is more advanced and examples of this prototyping technique include selective laser sintering [20], 3D

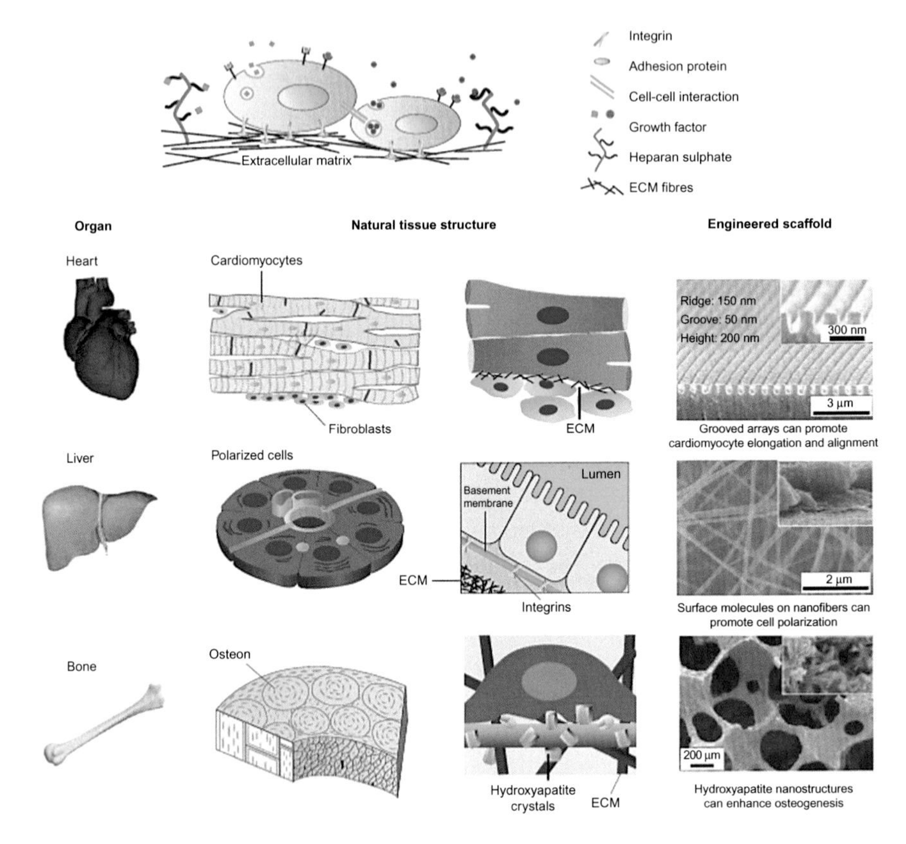

Figure 7.4 Examples of 3D scaffolds in tissue engineering from different human tissues.

printing [21], and lithography [22]. More recently, self-organized honeycomb porous structures using block copolymers [23] have been developed. The following section highlights the recent developments in scaffold fabrication by lithography and 3D printing and also elaborates on self-organization methods as well as self-assembly of peptides, specifically for the enhancement of cellular functioning in tissue engineering applications.

3D SCAFFOLD FABRICATION BY LITHOGRAPHY AND PRINTING TECHNIQUES

Polymer patterning of 3D surfaces in biomedical research to study cellular behavior and tissue engineering [24–26] has generated a great deal of interest worldwide. Because of this, a great deal of advancement has taken place in this technology recently. In particular, polymeric biomaterials and crosslinked HGs have found a wide range of applications in microdevices using various approaches. In the following recent developments in HG patterning using photolithography, dip-pen lithography, nanoimprinting, contact printing, solid-free form, robotic deposition, and their application in tissue engineering, are described.

Photolithography is one of the most well-known fabrication methods used to generate a 3D structure and a pattern using various molecular weights of polymeric materials [27–40]. Photolithographic patterns can be generated in polymer films and in monolayers [41].

The interference methods generate periodic patterns such as Bravais lattices [42].

Bioprinted tissues can help better predict and test whether a drug will be effective on people and at less cost, which is what researchers at the University of British Columbia Department of Electrical and Computer Engineering and spinoff Aspect Biosystems hope to prove.

Ultimately, this work could also lead to growing organs for human transplant.

A New 3D Printer Design for Tissues, Using a Microfluidic Chip

In July 2016, researchers from Organovo and Roche developed 3D printed human organs. Currently 3D printed human tissues are already playing a different yet similarly important role in medicine. Using 3D bioprinted tissue, researchers in the pharmaceutical industry are now able to test the effect of substances on "human" organs, but without the need for human volunteers. This allows researchers to perform much more radical experiments than would otherwise be possible, since there is no human life at stake.

Organovo, a specialist in 3D bioprinted human tissues, has been providing artificially fabricated tissues for pharmaceutical research for several years now, and a new study conducted by Organovo and Roche, a global healthcare company, has shown that Organovo's 3D printed human liver tissues can be used to successfully distinguish between the differing toxicity levels of multiple substances. This capability could help pharmaceutical companies identify substances that could potentially produce harmful effects on the human body.

Using Organovo's 3D bioprinted human liver tissues, the researchers were able to detect the toxicity of trovafloxacin, a third-generation antiinfective drug that was withdrawn from the market 1 year after being approved after a small number of patients died of liver failure. When this substance was initially approved, traditional testing methods were unable to identify its toxicity. However, researchers at Roche and Organovo were able to distinguish between the toxicity profiles of trovafloxacin and levofloxacin, a structurally related, but nontoxic compound. Organovo hopes that the success of these experiments could pave the way for its 3D printed liver tissues to be used—in conjunction with other methods, such as animal testing and 2D cell culture models—in the testing of future pharmaceutical products.

The histological effects of trovafloxacin on 3D bioprinted liver tissues. Loss of cellular adhesion is indicated by the single arrow; increased hepatocyte necrosis is indicated by the two arrowheads.

The 3D printed human liver tissues produced by Organovo are composed of patient-derived parenchymal (hepatocyte) and nonparenchymal (endothelial and hepatic stellate) cell populations, and are both spatially patterned and 3D. The 3D nature of the tissues enabled the Roche and Organovo researchers to identify distinct intercellular hepatocyte junctions, CD31 + endothelial networks, and desmin-positive, smooth muscle actin-negative quiescent stellates. The 3D printed tissues also maintained metabolically relevant levels of ATP and albumin, as well as drug-induced enzyme activity of cytochrome P450s for more than 4 weeks in culture—something that 2D cell cultures would be unable to achieve.

REFERENCES

[1] Wichterle O, Lim D. Nature 1960;185:117.

[2] Kwon IC, Bae YH, Kim SW. Nature 1991;354:291.

[3] Hoffman AS. Adv Drug Delivery Rev 2002;43:3.

[4] Shi D. Introduction to biomaterials. Beijing: World Scientific, Tsinghua University Press; 2006.

[5] Liu WG, Griffith M, Li F. J Mater Sci Mater Med 2008;19:3365.

[6] Yang F, Wang Y, Zhang Z, Hsu B, Jabs EW, Elisseeff JH. Bone 2008;43:55.

[7] Eljarrat-Binstock E, Orucov F, Frucht-Pery J, Pe'er J, Domb AJ. J Ocul Pharmacol Ther 2008;24:344.

[8] Khan F, Tare RS, Oreffo ROC, Bradley M. Angew Chem Int Ed 2009;48:978.

[9] Nuttelman CR, Rice MA, Rydholm AE, Salinas CN, Shah DN, Anseth KS. Prog Polym Sci 2008;33:167.

[10] Hennink WE, van Nostrum CF. Adv Drug Delivery Rev 2002;54:13.

[11] Kamath KR, Park K. Adv Drug Delivery Rev 1993;11:59.

[12] Hollister SJ. Nat Mater 2005;4:518.

[13] Curtis ASG, Wilkinson CDW. Biomaterials 1997;18:1573.

[14] Dalby MJ, Gadegaard N, Tare R, Andar A, Riehle MO, Herzyk P, et al. Nat Mater 2007;6:997.

[15] Asefnejad A, Khorasani MT, Behnamghader A, Farsadzadeh B, Bonakdar S. Int J Nanomed 2011;6:2375.

[16] Allaf RM, Rivero IV. J Mater Sci: Mater Med 2011;22:1843.

[17] Wang CZ, Ho ML, Chen WC, Chiu CC, Hung YL, Wang CK, et al. Mater Sci Eng, C 2011;31:1141.

[18] Saraf A, Baggett LS, Raphael RM, Kasper FK, Mikos AG. J Controlled Release 2010;143:95.

[19] Reverchon E, Cardea S, Rapuano C. J Supercrit Fluids 2008;45:365.

[20] Duan B, Wang M, Zhou WY, Cheung WL, Li ZY, Lu WW. Acta Biomater 2010;6:4495.

[21] Khan F, Ahmad SR. In: Ramalingam M, Ramakrishna S, Best S, editors. Biomaterials and stem cells in regenerative medicine. USA: CRC Press, Taylor & Francis; 2012. p. 101–21. ch. 5.

[22] Gates BD, Xu Q, Stewart M, Ryan D, Willson CG, Whitesides GM. Chem Rev 2005;105:1171.

[23] Widawski G, Rawiso M, Francois B. Nature 1994;369:387.

[24] Théry M, Racine V, Pépin A, Piel M, Chen Y, Sibarita J-B, et al. Nat Cell Biol 2005;7:947.

[25] Théry M, Racine V, Piel M, Pépin A, Dimitrov A, Chen Y, et al. Proc Natl Acad Sci USA 2006;103:19771.

[26] Seol YJ, Kang TY, Cho DW. Soft Matter 2012;8:1730.

[27] Beaulieu MR, Hendricks NR, Watkins JJ. ACS Photonics 2014;1:799.

[28] Park S, Kim D, Ko SY, Park JO, Akella S, Xu B, et al. Lab Chip 2014;14:1551.

[29] Comina G, Suska A, Filippini D. Lab Chip 2014;14:424.

[30] Voelkel R, Vogler U, Bramati A, Hennemeyer M, Zoberbier R, Voigt A, et al. Microsyst Technol 2014;20:1839.

[31] Muller M, Becher J, Schnabelrauch M, Zenobi-Wong M. J Visualized Exp 2012;67:e50632.

[32] Balowski JJ, Wang Y, Allbritton NL. Adv Mater 2013;25:4107.

[33] Lu Y, Chen S. Methods Mol Biol 2012;868:289.

[34] Liebschner MAK, Wettergreen M. Methods Mol Biol 2012;868:71.

[35] Revzin A, Tompkins RG, Toner M. Langmuir 2003;19:9855.

[36] Yamato M, Konno C, Utsumi M, Kikuchi A, Okano T. Biomaterials 2002;23:561.

[37] Karp JM, Yeo Y, Geng W, Cannizarro C, Yan K, Kohane DS, et al. Biomaterials 2006;27:4755.

[38] Menard E, Meitl MA, Sun Y, Park JU, Shir DJL, Nam YS, et al. Chem Rev 2007;107:1117.

[39] Albrecht DR, Tsang VL, Sah RL, Bhatia SN. Lab Chip 2005;5:111.

[40] Albrecht DR, Underhill GH, Wassermann TB, Sah RL, Bhatia SN. Nat Methods 2006;3:369.

[41] Menard E, Meitl MA, Sun Y, Park JU, Shir DJL, Nam Y-S, et al. Chem Rev 2007;107:1117.

[42] Moon JH, Ford J, Yang S. Polym Adv Technol 2006;17:83.

CHAPTER 8

Advances in Personalized Medicine: Bioprinted Tissues and Organs

Current tissue engineering methods can readily create materials, structures, and scaffolds that can support one or a few cell types in vitro, but in vivo many challenges exist that limit the efficacy of current, single-tissue designs. There will naturally be differences between living tissue and the implanted remedy, so researchers must take this into account and allow for proper integration without major mechanical and biochemical disparities. Furthermore, many injuries that would benefit from a tissue-engineered solution, such as osteochondral, ligament, and nerve damage, occur at the interface of two or more tissue types.

FDA FINAL GUIDANCE ON 3D PRINTING AND UNDERSTANDING THE SPECIFIC REGULATORY CHALLENGES

Overview

The flexibility of 3D printing allows designers to make changes easily without the need to set up additional equipment or tools. It also enables manufacturers to create devices matched to a patient's anatomy (patient-specific devices) or devices with very complex internal structures. These capabilities have sparked huge interest in 3D printing of medical devices and other products, including food, household items, and automotive parts (Fig. 8.1).

Medical devices produced by 3D printing include orthopedic and cranial implants, surgical instruments, dental restorations such as crowns, and external prosthetics. As of December 2015, the FDA has cleared more than 85 3D printed medical devices.

Due to its versatility, 3D printing has medical applications in medical devices regulated by the FDA's Center for Devices and Radiological Health (CDRH), biologics regulated by FDA's Center

Bioprinting. DOI: http://dx.doi.org/10.1016/B978-0-12-805369-0.00008-0

Figure 8.1 3D printed (left to right, top) models of a brain, heart, apophysis (elbow join in tan color), and lumbar vertebrae (in yellow color) printed on FlashForge Dreamer personal 3D printer.

for Biologics Evaluation and Research, and drugs regulated by FDA's Center for Drug Evaluation and Research.

> The FDA has developed this draft guidance to provide FDA's initial thinking on technical considerations specific to devices using additive manufacturing (AM), the broad category of manufacturing encompassing 3D printing. AM is a process that builds an object by iteratively building 2D layers and joining each to the layer below, allowing device manufacturers to rapidly alter designs

without the need for retooling and to create complex devices built as a single piece. Increased investment in the technology and its increased use in medical devices. The purpose of this guidance is to outline technical considerations associated with AM processes, and recommendations for testing and characterization for devices that include at least one AM fabrication step.

The considerations section describes the type of information that should be provided in premarket notification submissions [510(k)], premarket approval (PMA) applications, humanitarian device exemption (HDE) applications, de novo requests, and investigational device exemption (IDE) applications for an AM device. The type of premarket submission that is required for the AM device is determined by the regulatory classification of the device.

Point-of-care device manufacturing may raise additional technical considerations. The recommendations in this guidance should supplement any device-specific recommendations outlined in existing guidance documents or applicable FDA-recognized consensus standards. Biological, cellular, or tissue-based products manufactured using AM technology may necessitate additional regulatory and manufacturing process considerations and/or different regulatory pathways. The Agency encourages manufacturers to engage with the Center for Devices and Radiological Health (CDRH) and/or CBER through the presubmission process to obtain more detailed feedback for additively manufactured medical devices.

Types of medical devices are powder fusion, stereolithography, fused filament fabrication, and liquid-based extrusion. For medical devices, AM has the advantage of facilitating the creation of anatomically matched devices and surgical instrumentation by using a patient's own medical imaging.

While a variety of different types of materials can be additively manufactured, workshop participants noted that material control is an important aspect to ensure successful fabrication, and that final device performance is tied to the machine and postprinting processes. The interaction between the material and machine was also discussed in the process validation session, and the need for a robust process validation and acceptance protocol appropriate to the risk profile of the final device was identified. Printing parameters should be captured and validated for each machine/material combination. The discussion on the physical and mechanical assessment focused heavily on validation of the process and acceptance of devices and components after postprocessing. This draft guidance outlines technical aspects of an AM device that should be considered through the phases of development, production process, process validation, and final finished device testing. Similarly, not all considerations are expected to be addressed in premarket submissions of AM devices.

The first step is the design process, which can include a standard design with discrete prespecified sizes and models, or a patient-matched device designed from a patient's own medical images. Once the device design has been created, the software workflow phase begins, where the device design is further processed to prepare it for printing, printing parameters are optimized, and the build file is converted into a machine-ready format. Concurrently with this step, material controls are established for materials used in the printing of the device. After printing is complete, postprocessing of the built device or component (e.g., cleaning, annealing, postprinting machining, and sterilization) takes place. After postprocessing, the final finished device is ready for testing and characterization. The quality system should be applied across all of these processes (Fig. 8.2).

Some devices are specifically exempted by regulation from most quality service (QS) requirements. Manufacturers should refer to applicable regulations for their specific device type to determine what QS requirements apply. There are several AM technologies and different combinations of processing steps that can be used with each technology to build a device. A production flow diagram that identifies all critical steps involved in the manufacturing of the device, from the initial device design to the postprocessing of the final device, can help ensure product quality. In addition, a high-level summary of each critical manufacturing process step may be helpful in documenting the AM process used. The characterization of each process step should

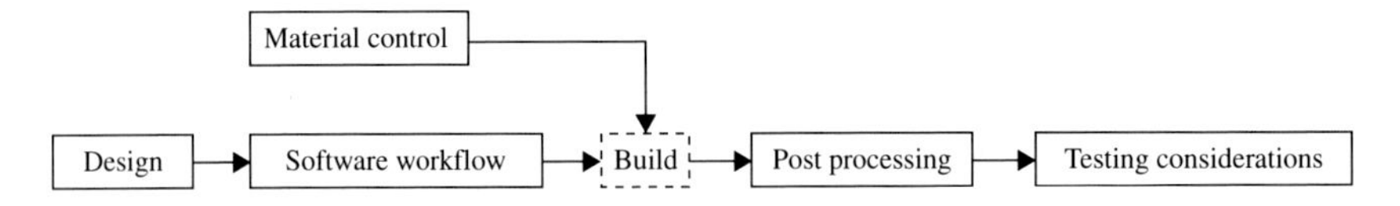

Figure 8.2 Flowchart of the additive manufacturing process.

include, but need not be limited to, a description of the process and identification of the process parameters and output specifications. Since processes that optimize one design parameter may influence information on processing steps should demonstrate the understanding of these trade-offs. Additionally, the cumulative effects of prior processes on the final finished device or component should be incorporated into the development of each process step and documented. It is important to use all reasonably obtainable knowledge about the specific machine's capabilities to ensure the manufacturing process outputs meet defined requirements [1–3]. Quantitative knowledge of the machine's capabilities and limitations can be gained through test builds, worst-case builds, or processes.

As with traditional manufacturing methods, design requirements drive the processes that can be used to reliably produce the device. Aspects of the "Global Harmonization Task Force Process Validation Guidance" may be helpful in developing process validation procedures.

1. Standard-Sized Device Design
   Standard-sized devices, or devices offered in preestablished discrete sizes, are often made by AM if they include features that are too complex to be made using other techniques. [8]ISO 14971 *Medical devices—Applications of risk management to medical devices* techniques. Specifically, one must compare the minimum possible feature size of the AM technique, in addition to the manufacturing tolerances of the machine, to the desired feature sizes of the final finished device. This is to ensure that devices and components of the desired dimensional specifications can be reliably built using the chosen additive technology. Dimensional specifications for the final device or component, as well as manufacturing tolerances of the machine, should be documented. Pixelation of features, where smooth edges become stepped, can lead to inaccuracies in final finished device dimensions.
2. Patient-Matched Device Design
   Patient-matched devices can be based on a standard-sized template model that is matched to a patient's anatomy. Note that while patient-matched or patient-specific devices are sometimes colloquially referred to as "customized" devices, they are not custom devices meeting the FD&C Act custom device exemption

requirements unless they comply with all of the criteria of section 520(b). For further information on custom device exemptions, refer to the Custom Device Exemptions guidance.
Patient-matched device designs may be modified either directly by clinical staff, the device manufacturer, or a third party in response to clinical inputs. Considerations for standard-sized devices are applicable for patient-matched devices.

Several factors may affect the fit of AM devices that use patient imaging to precisely control their size or

- any smoothing or image-processing algorithms that may alter the dimensions of the final device when compared to the reference anatomy,
- the clarity of anatomic landmarks used to match the device to the patient's anatomy.

If the device relies on anatomic features [1–6] that are not accurately imaged or are not consistent over time, then the final device may not fit the patient.

However, small changes in size or geometry may be difficult to identify during visual inspection of the device or through evaluation of patient imaging, and the mismatch may only be identified during device use. One should also consider the potential time constraints associated with producing an AM device based on the intended use of the device.

Many implantable devices and their patient-matched accessories depend on the patient's anatomy being identical to the recorded images in order for the device to function as intended. Therefore the labeled shelf-life of the device should account for the potential for time-dependent changes to the patient anatomy before the device is used.

Patient-matched devices are often made by altering the features of a standard-sized device for each patient within a predetermined range of device designs and size limits. One should also identify all medical devices and accessories that the design manipulation software is validated to work with.

File-Format Conversions

Errors in file conversion can negatively impact final finished device and component properties, such as dimensions and geometry.

Additionally, for patient-matched devices, all of the file conversion steps are typically performed for every device, whereas for a standard-sized device, most of the file-conversion steps would be performed once during the design phase. Therefore one should test all file-conversion steps with simulated worst-case scenarios to ensure expected performance, especially for patient-matched devices.

Digital Device Design

When a digital device design is finalized, additional preparatory processes are needed before the device can be additively manufactured [3]. This is commonly accomplished using build preparation software. These processes can generally be divided into four steps: (1) build volume placement, (2) add support material, (3) slice, and (4) create build paths.

Build Volume Placement

Placement and orientation of devices or components within the build volume is integral to individual device or component quality. The distance between each device or component can affect the material properties (e.g., poor consolidation or curing), surface finish, and ease of postprocessing. Orientation of each device or component can also impact its functional performance by affecting the anisotropic properties of the device or component.

Most AM techniques use a layer-wise printing process to fabricate components. For systems where layers are created by melting the material, the layer thickness can similarly influence the energy needed to create a uniform melt pool to enable bonding to the layer below.

Build Paths

The build path, the path traced by the energy or material delivery system (e.g., laser or extruder), can impact the quality of the final finished device or component. If more than one build path is used, each build path should be documented. It is also recommend that one assess whether differences in the build path significantly affect the performance of each component or device.

MACHINE PARAMETERS AND ENVIRONMENTAL CONDITIONS

Maintaining proper calibration and performing preventative maintenance have been identified as key factors to achieve low rejection rates of devices and components from an individual machine.

Environmental conditions within the build volume can also affect the part quality. Optimal settings and parameters for a single model of a machine can vary greatly when printing different devices or components. Modified by the device manufacturer and may have a significant impact on the device or component quality include, but are not limited to, build speed or beam speed and build path. Aspects of the "Global Harmonization Task Force Process Validation Guidance" also address installation qualification.

MATERIAL CONTROLS

In the AM process, the starting material may undergo significant physical and/or chemical changes. As such, the starting material can have a significant effect on the success of the build cycle, as well as on the properties of the final finished device.

- material supplier, and
- incoming material specifications and material certificates of analysis.

The specifications for incoming materials and test methods should be based on the AM technology used (i.e., material specifications will be different for powder-based vs stereolithography machines).

- if the material is a metal, metal alloy, or ceramic: chemical composition and purity,
- if the material is of animal origin, refer to: "Medical Devices Containing Materials Derived from Animal Sources (Except for In Vitro Diagnostic Devices)."

In addition, when any material specification is changed, the effect on the build process and the final device should be well understood and documented.

Postprocessing

Final device performance and material properties can be affected by postprocessing steps of AM (i.e., manufacturing steps occurring after the printing process). These steps could range from cleaning excess starting material from the device, through annealing the device to relieve residual stress, to final machining. All postprocessing steps should be documented and include a discussion of the effects of postprocessing on the materials used and the final device.

Process Validation and Acceptance Activities

Process Validation

Device quality, such as feature geometry, overall dimensions, material characteristics, and mechanical properties, are impacted by AM process parameters, process steps, and raw material properties, as described in the sections above. In addition, quality may vary when identical devices or components are built using different machines, even when the same machine model, parameters, process steps, and raw materials are used. Therefore knowledge of how the variability of each input parameter and processing step affects the final finished device or component is critical to ensuring part quality. Process validation must be performed to ensure and maintain quality for all devices and components built in a single build cycle, between build cycles, and between machines, where the results of a process (i.e., output specifications) cannot be fully verified by subsequent inspection and test [6]. Software must also be validated for its intended use according to an established protocol [6–8] (i.e., software workflow).

- in-process monitoring [9] of parameters such as:
- build-space environmental conditions (e.g., temperature, pressure),
- test coupon evaluation (see section V.E.4 Test Coupons).

Test methods used for process monitoring and control must be validated [10]. For example, analysis should be conducted to confirm that test coupons used are representative of the final finished device or component and representative of a certain area within the build volume.

A single failed component or device in a build cycle may not necessitate all devices or components within that build cycle to also be rejected. The criteria for determining whether to reject a single device or component, or the entire build, should be established before testing.

Changes to the manufacturing process or process deviations can trigger the need for revalidation, and these changes or deviations should be identified for each process.

- Certain software changes (e.g., change or update of build preparation software),
- Changes in material (e.g., supplier, incoming material specification, ratio of recycled powder) or material handling,
  - change in the spacing or orientation of devices or components in the build volume,
  - changes to the software workflow, and
  - changes to postprocessing steps or parameters.

Acceptance activities are integral to process control. Many AM technologies can produce more than one device or component simultaneously on different locations in the build volume. These devices or components can be copies of a single design or different designs. Some acceptance activities for individual devices or components can be performed through nondestructive evaluation.

Quality Data

For devices produced by AM, it is important to consider whether it is necessary to keep track of the location in the build volume where a device or component was built. This will depend on information obtained during process validation activities and design specifications. For example, if process validation demonstrated that quality is not affected by location in the build volume, it may not be necessary to be able to keep track of the build volume location for each device.

Device Testing Considerations

Depending on the intended use, risk profile, and classification and/or regulation for the device type. Not all considerations described will be applicable to a single device, given the variety of devices available [2,3,6–12].

In general, if the type of characterization or performance testing outlined in each of the subsections below is needed for a device made using non-AM techniques, the information should also be provided for an AM device of the same device type.

Device Description

AM facilitates the creation of intermediate and customized device sizes. Patient-matched devices are a good example of this application. Since these devices may not identify the range of dimensions for the device. Since each type of AM technology has different technical considerations, one should describe the type of AM technology used to build the device. Due to the generally complex geometry of AM devices, we recommend that critical features of the device be clearly described in the device description and identified in technical drawings. In the technical drawings of the device it is recommended that one identify components made using AM.

Mechanical Testing

The type of performance testing that should be conducted on a device made using AM is generally the same as that for a device manufactured using a traditional manufacturing method. Depending on the device type, these may include material property testing such as, but not limited to, modulus, yield strength, ultimate strength, creep/viscoelasticity, fatigue, and abrasive wear. Performance testing should be conducted on final finished devices subjected to all post-processing, cleaning, and sterilization steps or on coupons, if the coupon undergoes identical processing as the final finished device. Due to the nature of AM, devices will have an orientation (i.e., anisotropy) relative to the build direction and location within the build space. The orientation and build location can affect the final properties and should be considered when conducting device mechanical testing. Specifically, the build orientation (including worst-case orientation) of devices or components should be identified for each performance test.

Material Characterization

Since the AM process creates the final material or alters the starting material during the process, all materials involved in the manufacturing of the device should be identified. As noted in section V.C Material Controls, this information should include the source and purity of each material used. If material chemistry information in a device master file (MAF) will be referenced, one should include a right to reference letter from the MAF holder in the premarket submission [13]. One should also document the chemical composition of the final finished device.

Material Physical Properties

If the device is additively manufactured using a polymer, one must characterize the shore hardness and presence of voids or evidence of incomplete consolidation to ensure that the AM process is creating a device or component with uniform properties. For systems using a crystalline or semicrystalline material, crystallinity and crystalline morphology should be characterized to ensure that the AM process is not adversely altering the polymer structure and subsequently altering the performance (e.g., creep, material transparency) of the final device.

If the device is additively manufactured using an absorbable material, one should perform in vitro degradation testing using final finished devices or coupons. If coupons are used, they should be representative of the final finished device in terms of both processing and properties (e.g., surface-to-volume ratio, crystallinity). AM facilitates the creation of devices with complex geometries, such as engineered porosity, honeycomb structures, channels, and internal voids or cavities that cannot be produced by traditional manufacturing methods. Manufacturing material means any material or substance used in or used to facilitate the manufacturing process, a concomitant constituent, or a byproduct constituent produced during the manufacturing process that is present in or on the final finished device as a residue or impurity and not by design or intent of the manufacturer. There is also an increased risk of residual manufacturing material, such as excess starting material or support material, remaining on the final finished device. Since residual manufacturing material may negatively impact the performance of the device, one should describe how the cleaning process used ensures adequate removal of residual manufacturing materials as part of the cleaning validation process.

To evaluate the biocompatibility of the final finished device, one should do so as described in the guidance "Use of International Standard ISO-10993, 'Biological Evaluation of Medical Devices Part 1: Evaluation and Testing.'" If chemical device labeling should be developed in accordance with applicable regulations, device-specific guidance documents, and consensus standards. Since clinical staff, device manufacturers, or a designated third party might modify the design of each patient-matched device, additional labeling information is recommended for AM devices [1–5,14–16] that are patient-matched. Each patient-matched device should be marked or have accompanying physician labeling included in the packaging to identify the:

- patient identifier,
- final design iteration or version used to produce the device.

The expiration date for a patient-matched device may be driven by the patient imaging date or the design finalization date rather than the standard methods of determining device shelf-life.

ADDITIONAL RESOURCES

- FDA Voice: FDA Goes 3D—Describes 3D printing efforts at the FDA.
- Public Workshop—Additive Manufacturing of Medical Devices: An Interactive Discussion on the Technical Considerations of 3D Printing. Held in October 2014 to provide a forum for the FDA, medical device manufactures, AM companies, and academia to discuss technical challenges and solutions of 3D printing.
- How 3D Printers Work—A resource from the Department of Energy that includes descriptions of different types of printing processes.
- NIH 3D Print Exchange—Offers a unique set of models, learning resources, and tutorials to create and share 3D-printable models related to biomedical science. The goal of the project is to facilitate the application of 3D printing in the biosciences.
- American Society of the International Association for Testing and Materials (ASTM) International Committee F42 on Additive Manufacturing Technologies—This is a collaborative, consensus organization that has published standards and test methods for AM and 3D printing.
- America Makes—A public private partnership whose members, including the FDA, are working together to innovate and accelerate 3D printing to increase our nation's global manufacturing competitiveness.

What Is Currently Being Done in Biomedical 3D Printing

Some of the most incredible uses for 3D printing are developing within the medical field. Some of the following ways this futuristic technology is being developed for medical use might sound like a Michael Crichton novel, but are fast becoming reality.

Bioprinting is based on bioink, which is made of living cell structures. When a particular digital model is input, specific living tissue is printed and built up layer by cell layer. Bioprinting research is being developed to print different types of tissue, while 3D inkjet printing is being used to develop advanced medical devices and tools.

1. Organs
   While an entire organ has yet to be successfully printed for practical surgical use, scientists and researchers have successfully printed kidney cells, sheets of cardiac tissue that beat like a real heart, and the foundations of a human liver, among many other organ tissues [1–7]. While printing out an entire human organ for transplant may still be at least a decade away, medical researchers and scientists are well on their way to making this a reality.
2. 3D Printed Stem Cells
   Stem cells have amazing regenerative properties already—they can reproduce many different kinds of human tissue. Today, stem cells are being bioprinted in several university research labs, such as the Heriot-Watt University of Edinburgh. Stem-cell printing was the precursor to printing other kinds of tissues, and could eventually lead to printing cells directly into parts of the body.
3. Skin
   Imagine the uses that printing skin grafts could do for burn victims, skin cancer patients, and other kinds of afflictions and diseases that affect the epidermis. Medical engineers in Germany have been developing skin cell bioprinting since 2010 [1–3,6], and researcher James Yoo from Wake Forest Institute is developing skin graft printing that can be applied directly onto burn victims.
4. Bone and Cartilage: 3D printing on the bone
   Hod Lipson, a Cornell engineer, prototyped tissue bioprinting for cartilage within the past few years. Though Lipson has yet to bioprint a meniscus that can withstand the kind of pressure and pounding that a real one can, he and other engineers are well on their way to understanding how to apply these properties. Additionally, the same group from Germany who bioprinted stem cells is also working toward the same results for bioprinting bone and others parts of the skeletal system.
5. Surgical Tools
   Bioengineering students from the University of British Columbia recently won a prestigious award for their engineering and 3D printing of a new and extremely effective type of surgical smoke evacuator. Other surgical tools that have been 3D printed include forceps, hemostats, scalpel handles, and clamps—and best of all, they come out of the printer sterile and cost a tenth as much as the stainless-steel equivalent.

6. 3D Printing in Cancer Research
   In the same way that tissue and types of organ cells are being printed and studied, disease cells and cancer cells are also being bioprinted, in order to more effectively and systematically study how tumors grow and develop. Such medical engineering would allow for better drug testing, cancer cell analyzing, and therapy development. With developments in 3D and bioprinting, it may even be a possibility within our lifetime that a cure for cancer is discovered.
7. Heart and Blood Vessels
   Another German institute has created blood vessels using artificial biological cells, a 3D inkjet printer and a laser to mold them into shape. Likewise, researchers at the University of Rostock in Germany, Harvard Medical Institute, and the University of Sydney are developing methods of heart repair, or types of a heart patch, made with 3D printed cells.

The human cell heart patches have gone through successful testing on rats, and have also included development of artificial cardiac tissues that successfully mimic the mechanical and biological properties of a real human heart.

There are plenty of other developments being made with 3D and bioprinting, but one of the biggest obstacles is finding software that is advanced or sophisticated enough to meet the challenge of creating the blueprint. While creating the blueprint for an ash tray, and subsequently producing it via 3D printing is a fairly simple and quick process, there is no equivalent for creating digital models of a liver or heart at this point.

Technical and Regulatory Challenges

As usual, when technology is running so fast in a way that has been defined as garage-science, the legislature is in trouble chasing the different fast changes. If one may dream, e.g., of a 3D printer for building prosthesis for human livings, according to most strict interpretation of the international law, we need not only to guarantee the safeness of the whole production process, but also ensure sanitary standards that we normally find in hospitals and biomedical manufacturing environments. In the case presented by Zopf et al. [1,2,4,5,14–16], we notice also that, before doing the surgery, the clinicians need to have an emergency clearance from the Food and Drug Administration (FDA), since the polycaprolactone biopolymer does not have consensus to be used by the FDA. The production of medical devices inside the United States is strictly regulated by the FDA. So, before entering the marketing stage, a medical device needs to be certified by the FDA. However, for a medical device that does not appear in the FDA medical device database [17–19] the use is possible in some situations [17–19]. In one case, even though the procedure and the material was not intended to be used on humans, the surgery could take place because it was considered as a compassionate cure. Outside this context, i.e., a life or death situation, both the EU and US have strict regulations that assure the safeness of the product itself both for patient and clinicians. In the above cited case the problem was somehow bypassed by the fact that the nursing process took place in US territory under the auspices of the same agreed regulation.

This leads to a potential interesting law problem, as usual when dealing with remote assistance. If we imagine a remote extrusion 3D printer that uses a polymer to build a splint in a region outside the United States, what is the best way to assure the patient the safety of the process and to reduce at minimum the risk of rejection by the patient? And upon this, in case something goes wrong, who is the actor responsible and for what is he/she responsible? As in most of the latest technology breakthroughs, for the moment, the technology wave innovation is leading us to new and unseen possibilities, and not only for the western countries. Once the tide lowers a little, regulation should be introduced to manage the issues arising from the adoption of this new technology.

The time is approaching, because in February 2014, key patents that currently prevent competition in the market for the most advanced and functional 3D printers will expire. When this happens, we will most likely see a huge drop in the price of these devices. This happened when the key patents expired on a more primitive form of 3D printing, known as fused deposition modeling (FDM). The result was an explosion of open-source FDM printers that eventually led to iconic home and hobbyist 3D printer manufacturers. When the medical use of 3D printers becomes widely spread, it is time to initiate conversations about the practitioner's work with 3D printers. Also, systematic evaluations of the use of 3D printers will be beneficial to the area.

New Technologies in Development

This means a fully successful implant must simultaneously support the growth of different cell types and tissues, each with specific mechanical properties, chemical gradients, cell populations, and specific geometric constraints incorporated within the scaffold design. The myriad of complex design constraints limits the effectiveness of many current methods, especially when attempting to regenerate clinically relevant injuries, organs, and other complex tissues and tissue interfaces. To address the inherent limitations and requirements posed by the are increasingly being utilized. Nanobiomaterials have clearly played an integral role in tissue engineering, and will continue to be an important design consideration for future work. The beauty of incorporating nanomaterials into tissue-engineered constructs is the versatility they contribute almost intrinsically. Many nanomaterials add similarly vast improvements to the constructs they are incorporated within. The cutting edge of tissue engineering and regenerative medicine research is moving toward customizing therapies to individual patients and individual situations. In order for nanomaterials to be integrated into a patient-specific scaffold, manufacturing techniques need to be employed to allow for further micro- and macroscale customization; exactly what 3D printing excels at. Now, with the introduction of 3D printing in the tissue engineering field, researchers can begin to experience the benefits of having truly unique solutions to problems not easily solved with traditional fabrication techniques. A robust hobbyist community and open-source movement, RepRap, has been driving down the cost of implementation of many 3D printing technologies, making them available for researchers and clinicians in the research community, and thus 3D bioprinting has been advancing at an exciting pace. The successful implementation of a scaffold with complex requirements, including the utilization of multiple disparate materials and several nanomaterial constituents into a patient-specific geometry with highly defined internal microgeometry, has become surmountable. By incorporating multiple cell types, biomaterials, and nanomaterials in specific, biomimetic geometries, tissue engineers can expect to develop truly revolutionary medical devices, therapies, and treatments, and potentially usher in a new age of organ replacement.

In any transplant or surgery, there is always the risk of the body rejecting the organ or cells. This can even occur when tissue from one area of the body is put into another area of the body. The organ (or piece of tissue) also has to have time to integrate into the body after the implant. Since the technology for 3D bioprinting is so new, doctors and engineers have not even gotten to this point yet, but it's important to recognize these risks well in advance.

REFERENCES

[1] Symes M, Kitson P, Yan J, Richmond C, Cooper G, Bowman R, et al. Integrated 3d-printed reactionware for chemical synthesis and analysis. Nat Chem 2012;4:349–54.

[2] Zopf DA, Hollister SJ, Nelson ME, Ohye RG, Green GE. Bioresorbable airway splint created with a three dimensional printer. N Engl J Med 2013;368(21):2043–5.

[3] Carden KA, Boiselle PM, Waltz DA, Ernst A. Tracheomalacia and tracheobronchomalacia in children and adults: an in-depth review. CHEST J 2005;127(3):984–1005.

[4] Ilievski F, Mazzeo AD, Shepherd RF, Chen X, White-sides GM. Soft robotics for chemists. Angew Chem Inter Ed 2011;50(8):1890–5.

[5] Hasegawa T, Nakashima K, Omatsu F, Ikuta K. Multi-directional micro-switching valve chip with rotary mechanism. Sens Actuat A Phys 2008;143(2):390–8.

[6] Buys P, Dasgupta S, Thomas TS, Wheeler D. Determinants of a digital divide in sub-Saharan Africa: a spatial econometric analysis of cell phone coverage. World Dev 2009;37(9):1494–505.

[7] Zurovac D, Talisuna A, Snow R. Mobile phone text messaging: tool for malaria control in Africa. PLoS Med 2012;9(2) [Online]. Available from: http://dx.doi.org/10.1371/journal.pmed.1001176.

[8] Tam M, Laycock S, Bell D, Chojnowski A. 3D printout of a dicom file to aid surgical planning in a 6 year old patient with a large scapular osteochondroma complicating congenital diaphyseal aclasia. J Radiol Case Rep 2012;6(1).

[9] Bidgood WD, Horii SC, Prior FW, Van Syckle DE. Understanding and using DICOM, the data interchange standard for biomedical imaging. J Am Med Inform Assoc 1997;4(3):199–212.

[10] Fedorovich NE, Alblas J, Hennink WE, Oner FC, Dhert WJ. Organ printing: the future of bone regeneration? Trends Biotechnol 2011;29(12):601–6.

[11] Brooker AF, Bowerman JW, Robinson RA, Riley LH. Ectopic ossification following total hip replacement incidence and a method of classification. J Bone Joint Surg 1973;55(8):1629–32.

[12] Orlando G, Baptista P, Birchall M, De Coppi P, Farney A, Guimaraes-Souza NK, et al. Re-generative medicine as applied to solid organ transplantation: current status and future challenges. Transpl Inter 2011;24(3):223–32.

[13] TED. (2012, Nov.) Tedtalks official web site. [Online]. Available: http://www.ted.com/.

[14] Robson AJ. Complex evolutionary systems and the red queen*. Econ J 2005;115(504):F211–24.

[15] Ahn BY, Duoss EB, Motala MJ, Guo X, Park S-I, Xiong Y, et al. Omnidirectional printing of flexible, stretchable, and spanning silver microelectrodes. Science 2009;323(5921):1590–3.

[16] Therriault D, White SR, Lewis JA. Chaotic mixing in three-dimensional microvascular networks fabricated by direct-write assembly. Nat Mater 2003;vol. 2:265–71.

[17] Mannoor MS, Jiang Z, James T, Kong YL, Malatesta KA, Soboyejo WO, et al. 3d printed bionic ears. Nano Lett 2013;13(6):2634–9.

[18] Thingiverse. Thingiverse official web-site; 2013, Oct. [Online]. Available from: http://www.thingiverse.com/.

[19] Mota C. The rise of personal fabrication. In Proceedings of the 8th ACM conference on Creativity and cognition, ser. C&C '11. ACM, 2011, pp. 279–288.

INDEX

Note: Page numbers followed by "*f*" and "*t*" refer to figures and tables, respectively.

Printed in the United States
By Bookmasters